Prologue to
Super Quantum Mechanics

Prologue to
Super Quantum Mechanics

Something is Rotten in the State of Quantum Mechanics

Victor Vaguine

ConsReality™ Press

ISBN: 978-1-936795-10-9
Library of Congress Control Number: 2011940874

Published in the United States of America
ConsReality™ Press.
Dallas, Texas
www.consreality.com

Publisher's Cataloging-in-Publication data

Vaguine, Victor Aleksey.
    Prologue to super quantum mechanics / Victor Vaguine.
    p. cm.
    Includes index.
    ISBN 978-1-936795-10-9
1. Quantum theory. 2. Physics--Philosophy. 3. Cosmology. I. Title.

QC174.12 .V34 2012
530.12 --dc22   2011940874

Editing by Anne Smith.
Cover illustration Copyright © 2010 by ConsReality, Inc.
Cover design by Andre Vaguine / ConsReality, Inc.
Illustrations by Andre Vaguine / ConsReality, Inc.
Book design and production by Andre Vaguine / ConsReality, Inc.
Author photograph by Dwayne Horii.

Edition 1.0

ConsReality is a trademark of ConsReality, Inc.

To my wife Vera, son Andre and daughter Galena.

Acknowledgments

My sincere appreciation to Anne Smith for her professional work in editing the book.

My profound thanks to my son Andre Vaguine for his enduring and dedicated support and his active involvement in many areas of work. He was instrumental in bringing the book to publication.

Deep gratitude to my wife Vera Vaguine for her efforts working with files, typing and re-typing, and organizing the references.

My sincere appreciation to my daughter Galena Miller-Horii for putting final touches to the manuscript.

I want to thank Tina Vaguine and Dwayne Horii for their understanding and support.

I am also encouraged and delighted seeing my grandchildren Nicole, Danielle, Veronica and Ross expressing intellectual curiosity by reading various chapters of the book and supplying surprisingly good help in reviewing.

Contents

List of Illustrations

Introduction

This is a prologue to my other two books, *Conceptual and Philosophical Foundations of Super Quantum Mechanics*[1] and *Ontology of ConsReality*[2].

Surprisingly, the microscopic quantum world is intrinsically interconnected with the origin of our Universe, especially at the pre-Big Bang state. In contrast to a materialist view, which states that the Universe originated from primordial chaos, in this book I take exactly the opposite view: our Universe originated from the "absolutely" organized pre-Big Bang state called the Cosmic Seed. The pre-Big Bang state includes "absolute" entanglement and "absolute" zero temperature, among other "absolute" properties.

At the 1927 Solvay conference, Niels Bohr and Werner Heisenberg, as principal architects of the Copenhagen school, declared that the theory of quantum mechanics was complete and required no more debates. Einstein strongly disagreed but offered no alternative. He did not realize that as a local realist he was not in a position to offer an alternative.

For more than eight decades, quantum mechanics was held hostage by the Copenhagen interpretation with its quantum positivism. Murray Gell-Mann, Nobel laureate stated, "Niels Bohr brainwashed a whole generation of physicists into believing that the problem had been solved."[3] All this time, countless enigmas, mysteries, and paradoxes have been strong indicators that the quantum mechanics paradigm is wrong.

Einstein never accepted quantum mechanics as a complete theory. He declared that as a statistical science, quantum mechanics is compelling but it is not a complete theory for it does not include a description of an individual quantum system. He said, "Every element of physical reality must have a counterpart in the physical theory." Einstein was unable to offer an alternative theory—a deep theory, for he refused to accept non-locality as a property of objective reality. He stayed firmly on the position of local realism. To date, quantum mechanics remains a probabilistic and epistemic science with no ontology.

In quantum literature, you read again and again that quantum mechanics is the most successful science and has never been experimentally proven wrong. This statement became a cliché. But beware! Before continuing, I want to tell you a story; picture the following hypothetical scenario.

The time: the epoch in our past when Flat Earth mentality was the predominant worldview.

The place: a beautiful beach at seaside.

It's a beautiful day. The sun is shining and the bright blue sky is filled with white clouds. The sea is quiet, small waves sparkling under the sun. On the beach there are five men, sitting comfortably, talking, drinking wine, and eating fruit. They are sages, each well-known in their time. Some traveled a long way to get here. They have a very important topic to discuss—what holds up the Earth? A ship with square sails is slowly heading to the open sea. A stranger passing by sees the group and recognizes some of the sages. They are so famous. The stranger is overwhelmed by the opportunity to meet them. He approaches the group and greets the sages. He then asks the question, "Are you sure that the Earth is flat?" The answer comes, "Absolutely sure! So far, no one has been able to prove otherwise." Meanwhile, the sailing ship is disappearing from view not only horizontally but also vertically. No one pays much attention to it.

In our time we have our own cliché—no one has ever proven quantum mechanics wrong experimentally. In my view, the reason is that our mind is blocked by quantum positivism. Is there a way to demonstrate experimentally the limits beyond which quantum mechanics is no longer valid? Yes, there is.

In contrast to epistemic quantum mechanics, Super Quantum Mechanics is an ontological science. The philosophical foundation of Super Quantum Mechanics is non-local realism. In its turn, non-local realism is a subset of my more general philosophy called *ConsReality* as presented in Reference 2 of the Introduction. The centerpiece of Super Quantum Mechanics is an elementary quantum entity. Each quantum entity is tangible, with all its physical attributes at all times, can be visualized and is not hanging in limbo. Each quantum entity has a precise position and a precise momentum.

In our time, experimental physicists can manipulate and work with individual quantum entities such as a single photon or a single electron. Those entities are not abstract. Quantum positivism prevents scientists from even raising the issue of visualization. Bohr declared, "There is no quantum world. There is only an abstract quantum mechanical description." On the contrary, humans have an inherent ability to visualize the quantum world and even the next, deeper world. Objective reality in our Universe is three-dimensional on all levels. Without visualization, humans would not be able to advance science and evolve toward a higher level of intelligence. But we have to give Bohr a break. In his time, several decades ago, scientific discoveries in the microscopic quantum world were overwhelming.

Do we actually know what a photon looks like? We know photon's physical parameters, such as energy, wavelength, momentum, spin and polarization, but these parameters are abstract numbers. We do not know the ontology of photon and how it looks in three-dimensional space. In quantum literature, photon is presented as a wave packet.

What about electron? What is a three-dimensional image of electron? Is it a fast spinning charged ball? This model has been tried many times. It does not work. To produce the required magnetic moment, spinning must be faster than the velocity of light. In addition, repulsive electrostatic forces inside the tiny ball are so huge that the electron would be instantly blown apart. Furthermore, quantum mechanics does not allow the electron to have both a precise position and momentum because of the uncertainty principle. This is what highly-respected British scientist Stephen Hawking and his co-author Leonard Mlodinow said: "Quantum physics tells us that nothing is ever located at a definite point because if it were, the uncertainty in momentum would have to be infinite. In fact,

15

according to quantum physics, each particle has some probability of being found anywhere in the universe."[4]

I want to challenge this. After all, I am not a quantum positivist. Consider a single electron. We do not need high energy in the Tera range (10^{15} electron volts); energy of 100 electron volts would be sufficient. We let our electron travel against a uniform electrostatic field. At a certain moment, the electron would stop and reverse its direction. In that particular moment, the electron would have a precise position and zero momentum, in violation of the principle of uncertainty. Who said that the principle of uncertainty is absolute? But according to John Wheeler: elementary phenomenon is not a real phenomenon until is an observed phenomenon. Wheeler is a perfect example of quantum positivism mindset.

Ontology of both photon and electron are solved by Super Quantum Mechanics and described in Reference 2 of the Introduction. Photon and electron are beautiful quantum objects.

The forefront of fundamental theoretical physics has moved into a high energy field and super-gravity, leaving behind unfinished business.

The best theoretical physicists are busy with superstrings, a theory without visualization and without future. The best cosmologists are developing various multiverse theories which exhibit extreme materialist philosophy.

You will find in this book certain repetitiveness. It is intentional in order to counterbalance the layers of absurdity that have accumulated in scientific literature.

A new paradigm of Super Quantum Mechanics requires development of a new vocabulary, such as *ConsReality, intropy* (inverse of entropy), *Uni-Universe, exclusively unique, element, universal fine tuning, local fine tuning, cosmic seed, sharp definition, absolute entanglement, crossing point, time window, qualified locations, autopilot,* and more.

Begin your journey. It is my hope that you will enjoy the ride.

Part I.

Quantum Mechanics Issues

1

"Nobody Understands Quantum Mechanics"

Quantum mechanics represents an outstanding scientific achievement of the 20th Century. It is the result of immense efforts by many theoretical and experimental physicists, especially by such scientific giants of the 20th Century as Max Planck, Albert Einstein, Niels Bohr, Louis de Broglie, Paul Dirac, Max Born, Wolfgang Pauli, Erwin Schrödinger and Werner Heisenberg. But eight decades after its foundation, quantum mechanics remains enigmatic and mysterious, full of paradoxes. As Niels Bohr observed, "Anyone who is not shocked by quantum theory doesn't understand the first thing about it."

Quantum mechanics accurately describes fundamental particles, structures of atoms and molecules, and a variety of complex physical phenomena such as superconductivity, Bose-Einstein condensates, strong and weak nuclear forces, and nuclear processes inside stars. But it only describes them statistically. Quantum mechanics is a probabilistic theory and as such is accurate. So far, not a single one of its predictions has ever shown to be wrong. Why then do top scientists still express such uneasiness and reservations about quantum mechanics?

Albert Einstein never accepted quantum mechanics as a complete theory especially in the Copenhagen interpretation with its quantum positivism bias. He did accept quantum mechanics as a probabilistic theory relative to the assembly of quantum events but considered it as an incomplete theory not suitable for description of an individual elementary quantum event. Einstein said, "I thought a hundred times as much about the quantum problems as I have about general relativity theory."[1]

Murray Gell-Mann, a Nobel laureate in physics, noted, "Quantum mechanics [is] that mysterious, confusing discipline, which none of us really understands but which we know how to use. It works perfectly, as far as we can tell, in describing physical reality, but is a 'counter-intuitive discipline', as the social scientists would say. Quantum mechanics is not a theory, but rather a framework within which we believe any correct theory must fit."[2]

Richard Feynman, also a Nobel laureate in physics declared, "I can safely say nobody understands quantum mechanics."[3]

David Bohm said, "....quantum mechanics does not explain anything: it merely gives a formula for certain results.... Quantum mechanics is a calculus that enables you to predict statistical results. But it has no explanations, and Bohr emphasized that there was no explanation of any kind.... Quantum mechanics says that nature is unintelligible except as a calculus, that all you can do is to compute with the equations and operate your apparatus and compare the results."[4]

John Bell stated, "Nobody knows what quantum mechanics says about any particular situation. Quantum mechanics explains nothing."[5]

British scientist Roger Penrose in his Foreword to the book *Quo Vadis Quantum Mechanics?* said, "... quantum mechanics—undoubtedly one of the supreme intellectual achievements of 20th Century—is still full of deep mysteries, despite the theory having been with us now, essentially in its modern 'final' form, for some three-quarters of a century."[6]

Bohr, Einstein, Gell-Mann, Feynman, Bohm, Penrose and Bell are telling us: No one understands quantum mechanics. Quantum mechanics is full of enigmas. Quantum mechanics explains nothing.

How is it possible that great scientists are saying "nobody understands quantum mechanics"? Yet, ask an undergraduate student who is studying quantum mechanics whether he understands quantum mechanics and he would likely say "Of course, I do." The majority of physicists who practice quantum mechanics are also convinced that they understand quantum mechanics. That is what Mark Silverman says "...it is abuse of language and inaccurate representation of physics to suggest that at present time no one understands quantum mechanics."[7]

If you ask an undergraduate student whether he understands the principle of complementarity, the answer would be "Of course, I do." There is a general understanding of the principle of complementarity among quantum scientists. The principle was originated by Niels Bohr who presented it at the Conference in Como, Italy in 1927. The principle states that you cannot measure particle and wave properties in the same experimental setup. You either measure particle properties or wave properties, but not both.

The principle of complementarity remains a valid principle even today. Some say that Bohr was lucky; it was a lucky, educated guess on his part. Perhaps. Great scientists are always lucky somehow.

But the principle of complementarity is not an ontological principle. It describes but does not explain. In fact, quantum mechanics is not an ontological science. It is an epistemic science. It is not by chance that John Bell said "quantum mechanics explains nothing." It is not by chance that Einstein, the most important scientist of the 20th Century, could not accept quantum mechanics as a complete theory.

There are many simple and basic questions which just do not go away. These questions have been asked many times. Instead, what you hear are exhortations about the nature of space and time, the influence of present on past, the influence of an observer or human consciousness on the outcome of quantum measurement, the presentation of the universe as a single quantum system defined by its global wavefunction, Many Worlds interpretation with endless splitting of parallel universes, the description of an individual particle in two or several different places at the same time, and so on.

Here, I will ask only four questions and deal with the others later:

- How is it possible that an individual particle, such as electron (or photon), can go through two or even more slits simultaneously?

- How is it possible that an individual particle such as electron (or photon) can interfere with itself?

- How is it possible that a particle, such as electron (or photon), can be in two or more places at once?

- How is it possible that if you measure spin of one of the pair of entangled particles, your action instantaneously influences the spin orientation of the second particle, while both particles are traveling in opposite directions and separated by any distance?

The principal reason why prominent scientists have had and continue to have reservations relative to quantum mechanics is due to the fact that quantum mechanics has almost no ontological content and is mostly an epistemic theory.

The following principles form the foundation of quantum mechanics and all are epistemic:

- Principle of complementarity

- Principle of uncertainty

- Wave-particle duality

- Wavefunction (aside from its probabilistic interpretation)

- Collapse of wavefunction

- Non-locality and action at the distance

Of course, if you are satisfied with the epistemic nature of quantum mechanics, you can claim that you understand quantum mechanics. So far, its mathematical formalism has never failed: it predicts the result of every experiment with no exception, but only probabilistically. You can ask: what else does one need? The top scientists tell us: we need to see ontology. Epistemic quantum mechanics mathematical formalism cannot produce ontology. For this, one has to shift the paradigm in a new direction.

All those unresolved mysteries, enigmas and paradoxes in quantum mechanics originate from the absence of ontological content and an erroneous paradigm. This book and its companion book (see the Introduction) is an attempt to move quantum mechanics decisively toward ontology. It is a daunting and ambitious task, my friends, especially for someone who is strictly speaking "an outsider" to the scientific establishment; who is not a part of any scientific groups, and who after leaving CERN (European Center for Nuclear Research), Geneva, Switzerland, in 1971, has been operating either in established private industry or in various entrepreneurial start-up companies.

This book is a prologue toward ontology in quantum mechanics as the reader can find out for him- or herself. With the inclusion of ontology, I am introducing the next step in quantum mechanics: Super Quantum Mechanics.

2

Planck and Einstein Opened the Quantum Mechanics Curtain

In 1894, Max Planck, a professor at the University of Berlin, Germany, began his work on black body radiation. At that time, no one had been able to describe mathematically a distribution of radiation spectrum observed and measured experimentally, for example, from a small hole in a perfectly black box at a high temperature.

In 1900, Planck succeeded in obtaining the correct mathematical formula by reluctantly introducing a counterintuitive assumption that the radiation does not flow continuously as one would expect from a classical Maxwell electromagnetic theory. Planck postulated that electromagnetic energy was emitted in quantized form, in multiples of an elementary unit $E = hv$, where h is a Planck constant, as it is known today, and v is a frequency of the radiation. The name *quantum mechanics* comes from Planck's unit of energy which he referred to as *quantum* or *quanta*.

Initially Planck considered that quantization was only '...a purely formal assumption.' Planck tried to understand the meaning of his own creation—energy quanta. "My unavailing attempts to somehow re-integrate the action quantum into classical theory extended over several years and cost me much trouble." For his fundamental discovery Planck was awarded the Nobel Prize in Physics in 1918.

Einstein, in 1905, achieved further progress on quantization in connection with his theoretical work on photoelectric effect, a known phenomenon at the turn of the century. Scientists could experimentally observe photoelectric effect, the emission of electrons from the surface of certain metals, by shining light on the metal. Two peculiar features could be observed but no explanation was found. First, the light was distributed over the metal surface, while emitted electrons as particles were localized. Second, the kinetic energy of these "photoelectrons" was somehow fixed. A high-intensity light would knock out more electrons but their kinetic energy still remained unchanged. Einstein was able to explain the photoelectric effect by assuming that light consists of quanta (photons). He was the first who assigned both properties to quanta: wave and particle or, as it is known today, wave-particle duality. It was a fundamental achievement in quantum mechanics.

Einstein's photoelectric theory was not understood by the majority of physicists and was considered by some as "reckless." Only in 1923, after Compton experimentally demonstrated electron-photon scattering, was the wave-particle duality of photon accepted.

It was Einstein who brought a deep sense of "quanta" as a real, physical entity. For this reason the philosopher and historian of science Thomas Kuhn stated that Einstein, more than Planck, should be given credit for quantum theory.[8]

Today, both Planck and Einstein are considered the scientists who began, in my words, opening a curtain behind which quantum mechanics was hiding, leading to the discovery of the amazing and counterintuitive microscopic world of quanta.

Lord Kelvin in 1900, in his lecture to the Royal Institution all but declared physics complete, save for two small clouds in the otherwise blue sky of theoretical physics. One small cloud was transformed into the most successful physics theory—quantum mechanics, which however, even today cannot find its interpretational floor, and the second cloud grew into Einstein's special and general relativity. Just a few years after Lord Kelvin's famous statement about the end of physics, the landscape of physics was totally transformed.

3

Copenhagen Interpretation and Quantum Positivism

Quantum mechanics has many interpretations. Each interpretation has many versions or subdivisions. The fact that there are so many interpretations, including those that conflict among themselves, is a strong indicator that something is wrong with a quantum mechanics paradigm. The Copenhagen interpretation is widely recognized and considered the standard interpretation of quantum mechanics. In my view, however, the Copenhagen interpretation is "the worst except for all the others."

You can find many quantum mechanics books and scientific articles describing and analyzing various quantum mechanics interpretations. After studying all those seemingly endless versions of quantum mechanics interpretations, a physicist would be thoroughly confused. It is not by chance that many physicists are giving up on quantum mechanics interpretations, saying "philosophy is useless."

How can you judge which interpretation is worthwhile to consider and which is not?

From the position of Super Quantum Mechanics, if an interpretation does not include the collapse of wavefunction postulate, the interpretation is invalid. This guidance can help a reader to ignore the majority of the so-called interpretations. The collapse of wavefunction is a necessary but not sufficient criterion for the validity of an interpretation. Although the collapse of wavefunction has never been experimentally demonstrated, it was intuitively expressed by a number of early contributors to quantum mechanics. Einstein was the first (or one of the first) who noticed 'a peculiar mechanism' of wavefunction behavior in 1925 and expressed the same in his concluding comments at the 1927 Solvay conference. A few years later, the collapse of wavefunction was mathematically formalized by von Neumann.

In a companion book to this one (see Reference 1 of the Introduction), I present the collapse of wavefunction as a fundamental postulate and as a real, physical process. The collapse of wavefunction is ontologically transformed and its deep physical meaning explained.

After the Copenhagen interpretation, the two most analyzed and debated quantum mechanics interpretations are Hugh Everett's Many Worlds and Broglie-Bohm's Pilot Wave-Quantum Potential. In those interpretations, the wavefunction collapse is specifically eliminated. As a result, based on my criterion, both interpretations are not valid.

While Bohm's interpretation is a serious scientific attempt by an outstanding scientist, David Bohm, to find an alternative to the Copenhagen interpretation, in contrast, the Many Worlds interpretation proposed by Everett, and its derivative by David Deutsch, are absurd, pseudo-scientific, and speaking frankly, insane. The Many Worlds interpretation is popular with the general public but, in my opinion, the interpretation is so preposterous that I will attempt to expose its full absurdity in this book. Unquestionably, both Everett and Deutsch are talented scientists. The Many Worlds interpretation is an example that it is ideology, not a lack of talent, that breeds absurdity.

The Copenhagen interpretation of quantum mechanics was formulated during the foundation of quantum mechanics in 1920-1930. Principal contributors were Bohr and Heisenberg, supported by Pauli, Pascual Jordan, Max Born and von Neumann, who together became known as The Copenhagen Group. There existed strong opposition to the Copenhagen interpretation, principally from Einstein, and also from Schrödinger, Max Planck, and de Broglie. The Copenhagen Group prevailed. They had a power base, the Copenhagen Institute of Physics, plus strong leadership by Bohr and very loud voices in the scientific community. At the 1927 Solvay conference, Bohr and

Heisenberg declared that quantum mechanics theory is complete, done! And those who continued questioning it needed to accept the completion and move on. Einstein strongly disagreed, which only served to isolate him more and more from the quantum mechanics community.

The Copenhagen interpretation principally includes the following elements:

- Principle of complementarity by Bohr

- Uncertainty principle by Heisenberg

- Collapse of wavefunction, the concept mathematically formulated by von Neumann

In a narrow sense, quantum positivism is a philosophical basis of the Copenhagen interpretation, as it applies to quantum mechanics interpretation. Quantum positivism is a philosophy that is shaky on the concept of objective reality. On the one hand, quantum positivists recognize the objective reality of intrinsic properties in the description of quantum entities such as mass, charge, spin and magnetic moment. On the other hand, positivists deny the reality of existence associated with quantum entities that are not directly observed. Both Bohr and Heisenberg denied the physical reality of dynamical properties of quantum systems, such as position, velocity, momentum and energy, unless those systems are actually measured. Brian Greene's remarkable positivist description of quantum particles is well known to readers of his book *The Fabric of the Cosmos*, where he states "...particles hover in quantum limbo, in a fuzzy, amorphous, probabilistic mixture of all possibilities; only when measured is one definite outcome selected from the many."[9]

Werner Heisenberg stated, "In the experiments about atomic events we have to do with things and facts, with phenomena that are just as real as any phenomena in daily life. But the atoms or the elementary particles are not as real; they form a world of potentialities or possibilities rather than one of things or facts."[10] Pascual Jordan declared "Observations not only disturb what has to be measured, they produce it."[11]

In some interpretations an observer is elevated to a special position. It is the observer, by his actions or even through his human consciousness, who creates reality. One, however, wonders how our Universe was able to function before scientists with their measuring devices showed up.

In contrast, realists are convinced of the existence of dynamic properties of quantum entities and of the occurrence of quantum events independent of any measurements or any observations.

Bohr stated, "It is wrong to think that the task of physics is to find how nature is. Physics concerns what we can say about nature."[12] Bohr's statement confirms that quantum mechanics is an epistemic science.

His famous principle of complementarity states that one cannot measure wave-particle properties in a single experimental setup. It has to be two separate experiments: one for measurement of wave properties of a quantum entity and one for measurement of particle properties. The two experiments must be mutually exclusive. The principle of complementarity is epistemic; it describes but not explains.

The philosophy of quantum positivism affected such great scientific minds as Richard Feynman, John Wheeler and Stephen Hawking.

Bohr, Heisenberg, Pauli, Jordan, Max Born and von Neumann are outstanding scientists with massive contributions to the foundation of quantum mechanics; but as philosophers they are not in the same class with Einstein.

4

John Bell's Scientific Discovery

In the 1960s, the quantum physics community was oblivious to one important event. A CERN theoretical physicist, John Bell, published (as it turned out to be) a fundamental paper, "On the Einstein-Podolsky-Rosen Paradox"[13] in which he offered his Inequalities Theorem. It took several years for the physics community to awaken to the fundamental significance of John Bell's discovery. Prominent scientist Henry Stapp said about Bell's theorem: "...the most profound discovery of science."[14] And physicist Brian Josephson, Nobel Prize laureate, described Bell's theorem as the most important recent advance in physics.

Bell's theorem implied that quantum reality has non-local influences, expressed in the form of super-luminal connections, which exist between two entangled quantum systems that are separated by any distance.

To confirm the validity of Bell's theorem, several experiments were performed. The first experiment was performed in 1972 by Stuart Freedman and John Clauser from UC Berkeley, California,[15] then in 1976 by Edward Fry and Randall Thompson at Texas A&M,[16] and later was culminated by an experiment by Alain Aspect and his associates in Orsay, France in 1982,[17] who refined the method on the basis of the advanced technology.

In the wake of Bell's discovery, there have been many philosophical discussions and debates, which boiled down to a question of whether realism is a valid worldview. My answer is that it is absolutely valid but as a non-local realism.

In my opinion, future generations of physicists would view Bell's scientific discovery and follow-up experiments as a historic milestone and the beginning of a transition of quantum mechanics to a deeper stage that I call Super Quantum Mechanics (see Reference 1 of the Introduction).

As we compare Bell's theorem with the super string theory (I realize that this is an apples and oranges comparison), Bell's theorem is a modest exercise in mathematical formalism but fundamentally, it is incomparably more valuable. Furthermore, Bell's theorem made a prediction which was confirmed experimentally. The superstring theory has made no single prediction so far. As Peter Woit in his book Woit, author of the book *Not Even Wrong: The Failure of String Theory and the Search for Unity in Physical Law* says: "More than twenty years of intense research by thousands of the best scientists in the world producing tens of thousands of scientific papers has not led to a single testable experimental prediction of the [string] theory."[18]

5

Non-local Influences: Three Positions

If you want to select the most important enigmatic issue in quantum mechanics, it would be non-locality. Non-locality is built into quantum mechanics mathematical formalism and was noticed in the early days of its foundation. Einstein was the first to notice non-locality. In his concluding remarks at the 1927 Solvay conference, Einstein mentioned a "very peculiar mechanism of action at a distance" and its conflict with special relativity.

But what is non-locality? It is an instantaneous and non-local connection between spatially-separated, previously entangled particles. A simple example would be two quantum systems QS_1 and QS_2, which interacted (entangled) in the past and then were separated by a distance in space. Any action on system QS_1 produces instantaneous influence on system QS_2 regardless of the distance.

Einstein never accepted such a puzzling and mysterious phenomenon. It was a stumbling block Einstein was unable to overcome in his lifetime. As it turns out, non-locality is a real property of objective reality.

The interpretation of non-locality lay dormant for many decades since the foundation of quantum mechanics. And then in 1964, John Bell brought the issue back into sharp focus in his Inequality Theorem published in a scientific paper,[13] thus opening the venue for experimental verification. In the 1970-1980s, as technology became available, several groups of physicists performed experiments. As previously mentioned, among the scientists who experimentally demonstrated non-local influences were Stuart Freedman and John Clauser at UC Berkeley, Edward Fry and Randall Thompson at Texas A&M, Alain Aspect and his collaborators at Orsay, France, and others.

I will describe three positions on the issue of non-locality:

Position #1. Local or 'Naïve' Realism

Position #2. Quantum Positivism

Position #3. Non-local Realism

First, let us consider an example of a simplified non-locality experiment, as shown in Figure 1. Some technical details, such as collimators and coincidence electronics, are not included. Although the previously-mentioned experiments were performed with photons, here I will use non-relativistic electrons. Non-locality is a universal property of objective reality regardless whether one uses photons, electrons, protons, etc.

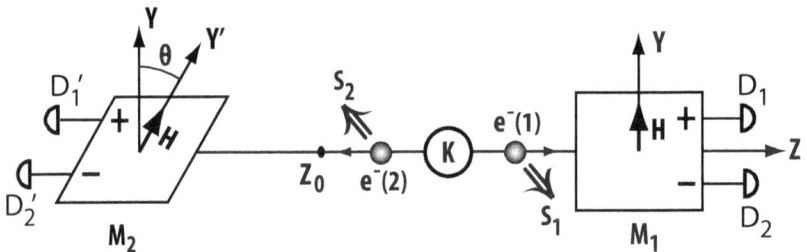

Figure 1: Non-locality experiment with electrons and Stern-Gerlach magnets

In this case, as shown in Figure 1, a source K produces pairs of entangled non-relativistic electrons which travel back to back along axis Z with their spins S_1 and S_2 aligned in opposite directions. From the point of quantum mechanics formalism, electron 1 and electron 2 represent a system with total spin equal to zero $(S_1 + S_2 = 0)$. It is assumed that electrons emitted from the source K are non-relativistic with an isotropic distribution of spin orientation. For spin measurements we use Stern-Gerlach magnets M_1 and M_2 positioned as shown in Figure 1. To simplify future explanation, I placed M_2 at a greater distance from the source K than M_1. The magnet M_1 is positioned for spin S_1 measurements along vertical axis Y, while M_2 is positioned for spin S_2 measurements along axis Y' at angle θ, as shown in Figure 1. The purpose of the experiment is to measure spin S_1 – spin S_2 correlation as a function of angle θ.

Position #1: Local or 'Naïve' Realism

In local—or what some call naïve—realism, there is no difference between entanglement and correlation. Both electrons of a given pair emerging from the source K have correlated spins aligned in opposite directions. Once a pair of electrons leaves the source, they travel apart independently back to back in free space along axis Z. During their travel their spin orientations are unchanged and stable because the law of spin conservation prevents any random fluctuations of spins orientations while traveling in free space. As fermion, electron has spin equal to one-half. Measurement of spin by magnet M_1 or magnet M_2 amounts to nothing more than to alignment of the orientation of electron magnetic moment either along or opposite to corresponding axis of measurement. A local realist assumes that interaction of electron 1 with magnet M_1 produces no influence on the spin orientation of electron 2 (no non-local influence of any kind). In such a case, a simple analysis should show that spin S_1 – spin S_2 correlation is a straight line A, as presented in Figure 2, which is in conflict with the experiment.

Local or naïve realism is a straightforward scientific and philosophical position and may remind you of classical physics. However, it is in conflict with quantum mechanics formalism and experimental results. Local realism is untenable as a scientific and philosophical position for the explanation of non-locality.

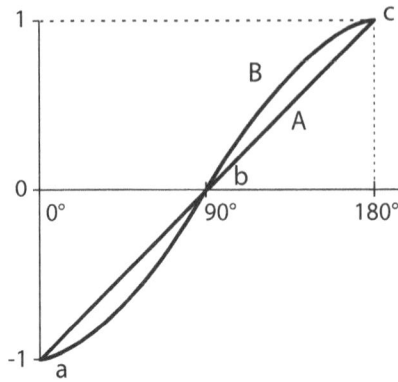

Figure 2: Spin S_1 – spin S_2 non-local correlations

Position #2: Quantum Positivism

According to quantum positivism, two entangled electrons represent one quantum system with total spin equal to zero. There are no pre-existing conditions. Positivist interpretation of quantum mechanics does not allow one to assume that the spins orientations of individual electrons 1 and 2 actually exist until they are measured. In his book *The Fabric of the Cosmos*, Brian Greene stated a positivism position in a clear way: "...particles hover in quantum limbo, in a fuzzy, amorphous, probabilistic mixture of all possibilities; only when measured is one definite outcome selected from the many."[9]

Brian Greene so precisely presents the positivist position that I would like to quote him again: "This sounds totally bizarre. But there is now overwhelming evidence for this so-called *quantum entanglement*. If two photons *[or electrons]* are entangled, the successful measurement of either photon's spin about one axis "forces" the other, distant photon to have the same spin about the same axis; the act of measuring one photon "compels" the other, possibly distant photon to snap out of the haze of probability and take on a definitive spin value—a value that precisely matches the spin of its distant companion. And that boggles the mind."[19] This is the majority position presented unambiguously by Brian Greene.

Quantum mechanics is a probabilistic science and describes correctly the assembly of quantum entities. Experimental results of spin S_1 – spin S_2 correlations confirmed John Bell's theory and are in agreement with quantum mechanics formalism. Statistical correlation of spin S_1 – spin S_2 is described by cosine θ function and is presented in this case by a curve B in Figure 2. But there is a difference. If a local realist accepts nothing but straight line A as spin-spin correlation, regardless of whether it is consistent with quantum mechanics formalism or not, then a quantum positivist accepts any correlation which quantum mechanics formalism provides. This is an example of a positivist quantum mechanics interpretation; it describes but explains nothing. It is enigmatic. Do positivists believe in objective reality? It is not clear to me... maybe yes, maybe not. But they do not believe it in the same sense as Einstein did. According to Einstein, a quantum entity possesses definite values of all its possible physical attributes regardless of the actions of observers or even their existence. Then, the question is who is right, local realists or quantum positivists? The answer is that both are right and wrong. Which means that both are wrong. Local realists are correct on insisting on the existence of objective reality independent of observers, but wrong in denying non-local influences as an objective property. Quantum positivists are wrong in denying realism or interpreting realism in fuzzy terms, but right in accepting quantum mechanics formalism with non-locality built in.

Then, there is a third position, which is the position of this author.

Position #3: Non-local Realism

Non-local realists are dissatisfied with the Copenhagen interpretation of quantum mechanics; they agree with Einstein that quantum mechanics is an incomplete theory. They consider that the centerpiece of future quantum mechanics is the elementary quantum entity or the quantum event. Of course, I am aware that many attempts have been made to convert probabilistic quantum mechanics into a theory that describes and explains elementary quantum events or entities. These earlier attempts failed, again and again, but I am willing to face the challenge. However, this is another story, outside of scope of this book, which is only a prologue to the greater discussion.

Let us review the data on spin-spin correlation as presented in Figure 2. According to non-local realism it is assumed that entangled electrons have pre-existing but unknown spin orientations and are traveling apart in free space along axis Z with their spins in existence as real attributes. During their travel, spin orientations are unchanged and opposite to each other with total spin equal to zero. Spin orientations are real, stable and not "a fuzzy, amorphous, probabilistic mixture of all possibilities." The entangled electrons with their spins are real objects existing independently of observers. But entangled electrons maintain a certain kind of non-local connection regardless of distance. Non-local connections are a property of objective reality and have to be recognized as such. The difference between cosine B and straight line A in Figure 2 represents the non-local influence. To illustrate this, consider a specific elementary quantum event. Electron 1 with a spin orientation at 45^0 relative to vertical axis Y enters magnet M_1 along axis Z. So-called spin measurement is nothing more than the alignment of electron magnetic moment along axis Y, which also means alignment of spin S_1 along the same axis. *Instantaneously, spin S_2 of electron 2 experiences a spin adjustment of 18^0*, which happened in location Z_0 while electron 2 is still traveling toward magnet M_2 (see Figure 1).

Because of the asymmetrical positions of M_1 and M_2 magnets relative to source K, a combination of electron 1 with magnet M_1 is considered a transmitter of non-local influences. What is most important is that non-local influences are not in conflict with special relativity, as it is explained in the companion book (see Reference 1 of the Introduction). In objective reality there are no paradoxes. Everything fits together. Paradoxes exist only in the minds of scientists or as a result of an undeveloped state of science. Science is an imperfect reflection of various aspects of objective reality. Sometimes, the resolution of some paradoxes must wait until science reaches a level when paradoxes can be explained. A paradox in physics is actually an opportunity for a young physicist to jump in and attempt to solve it. Once solved, the solution usually results in a scientific breakthrough and expansion of scientific territory.

Einstein stubbornly insisted on local realism and found himself in an untenable position. If he were still alive when John Bell derived his Inequality Theorem, Einstein might have considered the Theorem seriously. But he would definitely have been impressed with the non-locality experiments and would have found a solution to the quantum mechanics interpretation impasse. But in his lifetime, the technology was not available for such experiments. Einstein's mistake was that *for once in his lifetime, he deviated from his own philosophy: he prejudged what properties objective reality should or should not have.* The lesson is that objective reality has an 'infinite' supply of counterintuitive surprises and, ultimately, has an upper hand.

A non-local realist rejects quantum positivism as a philosophy that is incompatible with science. It rejects such statements as "particles hover in quantum limbo, in a fuzzy, amorphous, probabilistic mixture of all possibilities; only when measured is one definite outcome selected from the many."[9]

If you listen to a quantum positivist, you would learn that there is no such thing as the pre-existing orientation of spin until it is measured, regardless whether the particle with a spin is entangled or not. Contrary to quantum positivism, non-local realism states that although, as a rule, the pre-existing orientation of spin is unknown, it does exist. In fact, experiments using available technology have demonstrated the validity of this particular non-local realism position again and again.

For example, consider an experiment with vaporized silver atoms measured by a Stern-Gerlach magnet. Originally, this experiment was performed in 1922 by physicists Otto Stern and Walther Gerlach, and later described in detail by Daniel F. Styer in his book, *The Strange World of Quantum Mechanics.*[20] In the experiment, an electrical oven ejects vaporized silver atoms. The atoms enter the Stern-Gerlach magnet as shown in Figure 3, then exit from the magnet through a port (+) with spins up and a port (-) with spins down, also shown in Figure 3. Previously randomized unknown pre-existing spins are now thoroughly organized by the magnet in two beams: one with spin up and another with spin down. We can attach the second magnet to a port (+) as shown in Figure 3. With respect to the second magnet, silver atoms have known pre-existing conditions and in this case, the pre-existing conditions are confirmed by the output from the exiting ports (+) and (-). Only atoms with spin up exit from the port (+). No atoms with spin down exit from the port (-). The first and second magnets could be separated conceptually by any distance, and no 'hovering in quantum limbo' is expected. Furthermore, we could install a third magnet, a fourth magnet and so on to demonstrate redundantly that entering spins "do not hang in limbo, waiting for an experimenter's measurement to bring them into existence."[21] Figure 3 represents an example that defeats the quantum positivist statement that objective reality is produced by an observer's action.

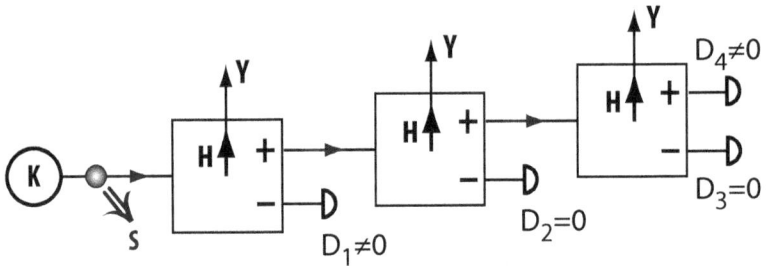

Figure 3: Stern-Gerlach experiment with silver atoms

Once again, we need to emphasize that silver atoms with random orientations of spins with pre-existing but unknown conditions enter the first magnet; silver atoms with organized spins up with pre-existing but known conditions enter the second magnet, and so on.

The same reasoning is applied to photons. To illustrate that, Figure 4 shows a simplified experimental layout. Source K produces a linearly-polarized beam of photons with a random

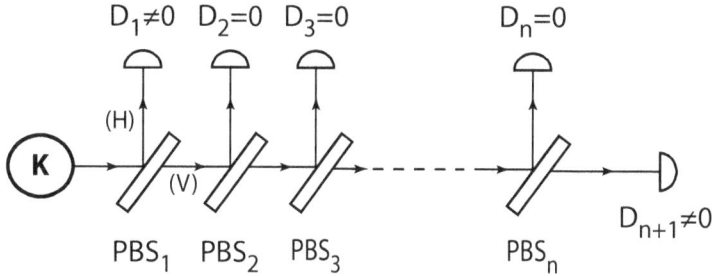

Figure 4: Quantum optics experiment with polarizing beam splitters

orientation of polarization planes representing pre-existing but unknown conditions. A polarizing beam splitter PBS_1 splits photons into two beams: a reflected beam with a horizontal plane of polarization (H) and a transmitted beam with a vertical plane of polarization (V). The transmitted beam is directed to a second polarizing beam splitter PBS_2 and then again, the resulting transmitted beam is directed to the third polarizing beam splitter PBS_3, and so on for a chain of additional beam splitters. The splitters can be separated by any arbitrary distance. Figure 4 shows several attached photon detectors D_1, D_2, ... D_n and D_{n+1}. Only detectors D_1 and detector D_{n+1} produce signals. The explanation here is the same as in the case of the experiment with silver atoms: unknown pre-existing conditions before splitter PBS_1 are transformed into known pre-existing conditions with respect to splitters PBS_2 through PBS_n. The experiment demonstrates that once a plane of polarization of a given photon is established, it remains stable in free space.

For more dramatic illustration, let us take an example of photons produced 380 thousand years after the Big Bang, when our Universe was transformed from a plasma state to a transparent state. Those photons were traveling for billions of years with a stable plane of polarization for linearly-polarized photons or a stable spin direction in the case of circularly-polarized photons.

At the present time, the technology has reached the level when single photon sources with "photon-on-demand" capability are available. The experiments shown in Figures 3 and 4 can be refined using these photons sources. Experimental physicists have acquired considerable experience with single photons. For them, photons with all their properties are real, quantifiable entities.

6

Quantum Mechanics and Enigmas

Quantum mechanics is full of enigmas. This fact is a strong indication that the quantum mechanics paradigm is wrong or at least has severe limitations and constrains. Enigmas arise when scientists attempt to explain some quantum phenomena beyond what the existing paradigm allows.

Quantum enigmas are described and discussed in many scientific articles and books, including the book by authors Bruce Rosenblum and Fred Kuttner *Quantum Enigma; Physics Encounters Consciousness.*[22]

Quantum mechanics is a probabilistic science and as such has never been proven experimentally wrong. But when we insist on determining trajectory, position and other dynamic variables of an individual quantum entity, we are stepping into a minefield of absurdities, enigmas and paradoxes. The details and specifics of an individual quantum entity cannot be determined on the basis of probabilistic calculations within the existing paradigm. If we still insist, then it would lead us to statements such as "photon takes one route, photon takes two routes" (John Wheeler).

We stay reasonably safe from a minefield of absurdities as long as we remain within the quantum mathematical formalism and probabilistic interpretation. We can take a specific quantum problem and, with some effort, sometimes very significant effort, calculate probabilities of occurrences of a specific quantum event. The difficulty arises when, in our attempt to solve a quantum problem, we are trying to extract its ontological content. In such a situation we have to expect to encounter one enigma after another. Many scientists have tried and some are still trying to arrive at ontological content within the existing paradigm, with no success.

Bohr warned over and over again not to attempt to analyze wave properties and particle properties in the same experimental setup because it would not work. An attempt to understand the ontological aspect of our mathematical description comes into conflict with Bohr's complementarity principle and Heisenberg's principle of uncertainty, one way or another. Within the existing quantum paradigm, there is no ontological solution. Since reality cannot be explained by the existing paradigm, the tendency is to reject reality as a naïve notion.

Enigmas in quantum mechanics are everywhere. Let us briefly review a few examples.

The Double-Slit Experiment

A stream of mono-energetic electrons is directed at two slits and forms an interference pattern on a detection screen. Electrons are separated from each other by sufficient distance to prevent mutual influence. In fact, this experiment can be considered as a one-at-a-time electron experiment repeated many times. The conclusion? *The electron interferes with itself.* Somehow, an electron passing through one slit is aware of the second slit. Or perhaps the electron passes through both slits at once? How can it be? How can a single electron participate either in constructive or destructive interference on the detection screen? Richard Feynman stated, "It is impossible to design any apparatus whatsoever to determine through which hole the electron passes that will not at the same time disturb the electron enough to destroy the interference pattern."[23] And Feynman goes on to say, "But the deep mystery is what I have described, and no one can go any deeper today."[24] For the two-slit experiment, Feynman added, "We chose to examine a phenomenon which is impossible, *absolutely* impossible, to explain in any classical way, and which has in it the heart of quantum mechanics. In reality, it contains the *only* mystery. We cannot make the mystery go away by 'explaining' how it works. We will just *tell* you how it works. In telling you how it works we will have told you about the basic peculiarities of all quantum mechanics."[25]

It is an enigma.

Mach-Zehnder Experiment

In accordance with Bohr's principle of complementarity, Figures 5A and 5B show the Mach-Zehnder experimental setup in two configurations: (A) one for the study of particle properties of photons with only one beam splitter BS_1 installed; and, (B) a second one for the study of wave properties of photons with two beam splitters BS_1 and BS_2 and an adjustable phase shifter. In the case of configuration A, a primary beam of mono-energetic photons from a source K is split into two equal secondary beams at the beam splitter BS_1. After being reflected by mirrors M_1 and M_2, the beams cross each other's paths at point C with no mutual interactions and are detected by detectors D_1 and D_2 showing statistically equal readings. Using a single photon source, we can reduce the primary beam intensity to the lowest possible level sending one photon at a time "on demand." Each individual photon has equal probability to be either transmitted through beam splitter BS_1 and follow a low path $BS_1 – M_1 - D_1$ (path L), or be reflected and continue along a high path $BS_1 – M_2 – D_2$ (path H). In this case, detectors D_1 and D_2 act as photon counters. Path H photons ("H-photons") will be counted by detector D_2 and path L photons ("L-photons") will be counted by detector D_1. As expected, we achieve statistically equal counting of photons between the two detectors. Each photon is detected by either detector D_1 or detector D_2, but never by both. Everything works as predicted by quantum mechanics. The experiment confirms that photons are particles. The photon is an indivisible particle; a partial photon has never been seen experimentally.

Now we revise our experimental setup in accordance with configuration (B) by installing the second beam splitter BS_2 and the adjustable phase shifter φ (see Figure 5B). The purpose of the adjustable phase shifter is to vary the optical length of path L relative to path H. At the phase shifter setting $\varphi = 0$, optical lengths of both paths H and L are equal and all photons are counted by D_1 and none by D_2. By varying the phase shifter we can demonstrate a counting reading for each detector as follows:

Detector D_1 $\qquad N_1 = N_0 \cos^2 \varphi$,

Detector D_2 $\qquad N_2 = N_0 \sin^2 \varphi$,

where $\varphi = \dfrac{2\pi}{\lambda} \Delta l$, Δl is adjustment in optical length in path L, λ is photon wavelength, and $N_0 = N_1 + N_2$.

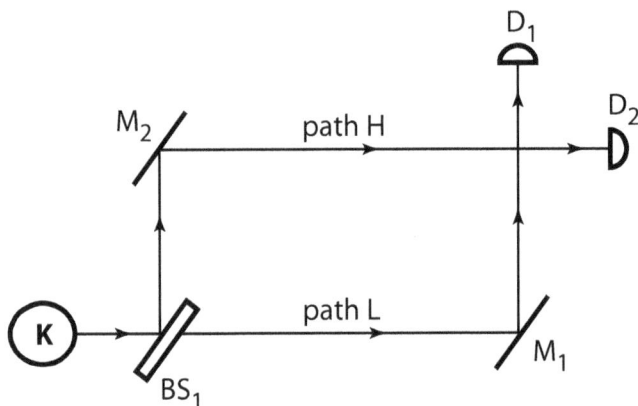

Figure 5A. Mach-Zehnder experiment in configuration A

Conclusion: Each photon behaves as a wave; it interferes with itself, and apparently travels through both paths H and L. As John Wheeler observed, "photon takes one route, photon takes both routes." Again, this is a quantum positivist interpretation. It cannot be true. It is an enigma.

We have experimentally demonstrated that the photon has both particle and wave properties. Two experimental setups A and B are complementary and consistent with Bohr's principle of complementarity, which describes but does not explain.

According to Bohr, one has to choose between "either tracing the path of the particle or observing interference effects." The principle of complementarity is epistemic; the ontological content is missing.

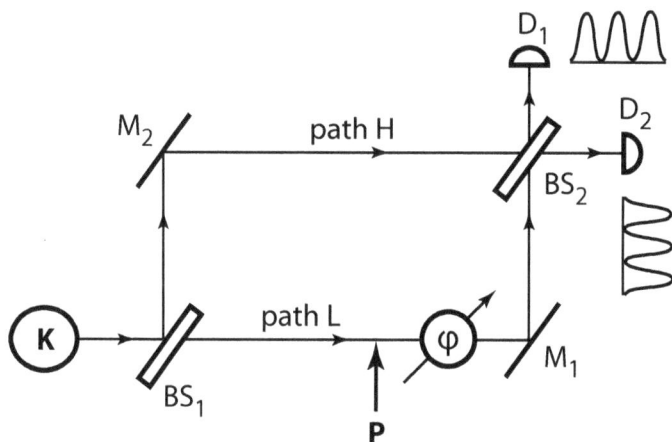

Figure 5B. Mach-Zehnder experiment in configuration B

There is something else in the experiment B. Suppose we insert probe P in path L (see Figure 5B). When probe P is inserted, it intercepts L-photons. The experiment shows that by inserting probe P, we "instantaneously" affect the behavior of H-photons: they are losing wave properties and behaving as particles by returning to statistically equal readings on detectors D_1 and D_2. We have some kind of *non-local* influence on H-photons in path H by acting at a distance by inserting probe P in path L. Einstein stubbornly rejected such "a spooky action at a distance."

It is an enigma.

The Collapse of the Wavefunction

The wavefunction is a solution of the Schrödinger wave equation. What is the meaning of the wavefunction? According to

Max Born, the wavefunction is an abstract mathematical entity describing the probabilistic aspect in quantum mechanics. Absolute square of the wavefunction amplitude determines a probability of finding a particle in a given space location.

However, there is another interpretation of the wavefunction presented for the first time by Einstein in his concluding comments at the 1927 Solvay conference. According to Einstein, the wavefunction in quantum mechanics defines the particle with all its attributes, such as momentum, position, charge, spin, etc.

Einstein presented a thought experiment where an electron passes through a pinhole and is detected on a semispherical screen. Einstein brought to attention a "peculiar mechanism" by which the electron wavefunction, distributed throughout the detection screen, instantaneously collapses into a location where the electron is detected. Einstein noted a conflict with special relativity.

So far, no one has experimentally detected a mysterious and instantaneous collapse of the wavefunction but the collapse is accepted as fact in the Copenhagen interpretation. In my view, the collapse is one of the strongest aspects of the Copenhagen interpretation.

It is also an enigma.

There is one more point, however. The wavefunction distribution is not limited to the detection screen. That fact causes some ridiculous positivist statements, such as, "So even if the chances of finding a given electron within double-slit apparatus are very high, there will always be some chance that it could be found instead on a far side of Alpha Centauri..."[26]

According to the quantum positivist position, instantaneous collapse of the wavefunction also means instantaneous appearance of a particle in a certain part of the distribution, regardless of distance.

Here is my position: You are wrong if you think you can detect the electron instantaneously in the Alpha Centauri location, and you are correct, if you think you have a chance to detect it in the Alpha Centauri location, provided you are willing to wait many, many years until the electron reaches the location.

Representing the majority view, Brian Greene stated, "After more than seven decades how or even whether the collapse of a probability waves really happens ... the collapse does not emerge from the mathematics of quantum theory; that has to be put in by hand, and there is no agreed-upon or experimentally justified way to do this. ...(H)ow is it possible that by finding an electron in your detector in New York City, you cause the electron's probability wave in the Andromeda galaxy to drop to zero instantaneously? ... How, in looser language, does the part of the probability wave in Andromeda, and everywhere else, "know" to drop to zero simultaneously?"[27]

It is an enigma.

Measurement in Quantum Mechanics

"Measurement" is a very unfortunate term in quantum mechanics. It implies that there is a human being who is doing the measuring. So, how was it possible for the Universe to function before humans came into existence? The term measurement is so entrenched in quantum mechanics literature that we have to live with it. According to Bohr, the measurement of a quantum entity is performed with a classic macroscopic device. Until it is observed or measured, a quantum object, such as an electron, does not exist. Between one measurement and the next, the quantum object has no existence outside of the abstract possibility of the wavefunction distribution. When an observation or measurement is made, the wavefunction collapses. Out of many "possible" states, the electron finds itself in the "actual" state.

Pascual Jordan went one step further. He declared that observation not only disturbs what has to be measured but creates it.

According to the Copenhagen interpretation, reality does not exist in the absence of observation. An electron with all its intrinsic properties does not exist at any place until a measurement or observation occurs. Neither does it exist before measurement or in between measurements. According to a majority, as presented by Brian Greene, "When they (particles) are not being observed or interacting with the environment, particle properties have a nebulous fuzzy existence characterized solely by a probability that one or another potentiality might be realized."[28]

According to Heisenberg, "atoms or the elementary particles themselves are not as real; they form a world of potentialities or possibilities rather than one of things or facts."[10]

So how is it possible that an electron with all its intrinsic properties, such as rest mass, electric charge, spin and magnetic moment, which are protected by physical conservation laws, does not exist until it is observed? It cannot be true. It is a fallacy of quantum positivism.

It is an enigma.

Nowadays, an experimental physicist working in the field of elementary particles ignores such philosophy as unhelpful. He/she knows that particles are real, with all their physical attributes, whether they are measured or not.

Let's look at the most important particle experiments performed on the Large Hadron Collider at the European Center for Nuclear Research, Geneva, Switzerland. Here, two beams of accelerated protons, each with the energy of 3.5 tera electron volts (tera is equal to 10^{12}), collide head to head producing a jet of secondary particles. Does a single proton-proton collision represent a measurement? Or is the registration of secondary particles by a detection system considered a measurement? But if the detection system was turned off and secondary particles were not detected, do they exist? Or if a jet of secondary particles misses the detection system partially or completely, do these secondary particles exist?

According to Brian Greene, a majority of scientists accept the Copenhagen interpretation of measurement/observation. "Particle properties, in this majority view, come into being when measurement forces them to..."[28] Again, a reader can see that a quantum positivist position creates confusion.

It is an enigma.

Sum over Histories à la Richard Feynman

Let us consider again a multiple-slit experiment with electrons. For a change, let us take ten slits instead of the usual double slit. According to the Copenhagen interpretation, one cannot talk about electron trajectory. On the contrary, says Richard Feynman, we should consider all possible paths of a single electron simultaneously through all ten slits toward the detection screen where an interference pattern is produced. Never mind that the electron is an indivisible particle. In Feynman's view, the paths in which the electron travels through one slit interfere with the paths in which it travels through other slits. How can a single electron travel *consequently* and *simultaneously* through all ten slits?

In conclusion: if one analyzes a movement of quantum particle from Point A to Point B, then, according to Feynman, all possible trajectories from A to B must be considered to obtain a correct result. It is called Feynman's Sum over Histories. Feynman states, "Anything that can happen must happen."

It is an enigma.

Sum over histories is a useful theoretical tool but as the interpretation it is a disaster.

Cosmic Experiment à la John Wheeler

My reader, as a sane person you know the difference between the past, the present and the future. But be vigilant – the brightest scientific minds tell you a different story. For example, here is the statement "Quantum physics tells us that no matter how thorough our observation of the present, the (un-observed) past, like the future, is indefinite and exists only as a spectrum of possibilities. The universe, according to quantum physics, has no single past, or history."[29]

Do not believe it. Rather than replace a wrong quantum paradigm with a correct one, the brightest scientific minds insist on absurdity. This is another example of mindset.

Now let's turn our attention to John Wheeler's proposed cosmic experiment. Photons emitted by a powerful quasar billions of light years ago are split into two paths and refocused toward Earth by the gravitational lensing of the intervening galaxy. If you measure an interference pattern here on Earth, photons behave as waves. If you detect them by a photon detector as particles, they behave as particles. By doing this, you are telling a photon how it should have behaved in the past: either as a wave or as a particle, the decision a photon took billions of years ago in anticipation of your action. You succeeded to influence a past decision, made billions years ago, *acting* from the present.

In such an interpretation, the relationship between the past, the present, and the future in quantum mechanics is an absurdity.

It is an enigma.

Linearly-Polarized Photon Quantum State

A wavefunction of quantum state of assembly of linearly-polarized photons $|LP\rangle$ is described as a superposition of right-hand circularly-polarized photons $|R\rangle$ and left-hand circularly-polarized photons $|L\rangle$:

$$|LP\rangle = \frac{1}{\sqrt{2}}(|L\rangle + |R\rangle)$$

The question is: how can one explain such superposition for a single linearly-polarized photon? The photon is indivisible. Are we to believe that a single linearly-polarized photon consists of one-half of a right-hand circularly-polarized photon and one-half of a left-hand circularly-polarized photon?

It is an enigma.

Spin

The Standard Model assumes that all fundamental particles, such as electrons, quarks, photons, neutrinos, muons, and gluons are point-like. They occupy no space. Then, how does one explain the angular momentum—spin—that they possess?

A point-like particle cannot have spin, but it does.

It is an enigma.

Electron in Energy Shell

Let us consider the dynamics of electron in the first energy shell in the simplest atom—hydrogen. The Schrödinger equation describes perfectly the probability distribution of finding electron in any location of the energy shell. Such distribution is confirmed experimentally. But then, how can we explain such a situation? One can offer three possible explanations, none of which is satisfactory:

1. Electron's mass and charge are dissolved within the energy shell in accordance with the probability distribution. Measurement of electron collapses the probability distribution and brings electron into existence (obviously, it is a positivist explanation).

2. Electron is located in a stationary position within the energy shell. Measurement of the assembly of such electrons confirms the probability distribution.

3. Electron orbits proton in a planetary mode but then, due to synchrotron radiation, electron loses energy and falls into proton.

The paradigm of quantum mechanics does not allow an explanation.

It is an enigma.

Non-Local Influences

In the case of two entangled particles, a spin direction of one particle is adjusted instantaneously at any distance defying special relativity when spin of the second particle is measured. The subject was discussed previously in the Chapter "Non-local Influences: Three Positions."

This is the biggest enigma of all.

In conclusion, the list of enigma examples can continue on and on. It is obvious that the paradigm of quantum mechanics is wrong. Scientists have been frustrated and giant efforts have been spent trying to resolve the issues. The progress was impeded by the ideology of quantum positivism and by Einstein's deviation from his own philosophy.

7

The Many Worlds Interpretation as Absurdity

In 1957, Hugh Everett, a student of John Wheeler, proposed the Many Worlds interpretation of quantum mechanics where the wavefunction is not required to collapse. According to Everett's theory, each and every potential quantum state is realized through its own separate universe. The Many Worlds interpretation requires an infinite number of "parallel universes" so that each and every quantum state, even states with extremely low probability, realizes its value in one of the universes. In each of the parallel universes there is a copy of you. John Wheeler was so excited about Everett's theory that he visited the famous Niels Bohr in Copenhagen to discuss the theory. Bohr was skeptical of the concept. In his professional life Bohr had heard many absurd ideas and theories but the Many Worlds interpretation was the top.

Eventually, John Wheeler abandoned the theory. As he explained "there is so much metaphysical baggage.... This is to make science into a kind of mysticism."[30]

Suppose you are a student with a passionate interest in quantum mechanics. All those mysteries and enigmas are thrilling. You are open-minded. You want to learn as much as possible. You want to accept the Many Worlds interpretation as a valid interpretation but you are getting resistance from your "inner voice." You ask yourself why so many prominent scientists keep working on further developing the Many Worlds interpretation. It must be a valid science if it is still actively pursued. You are in limbo – on the one hand you believe you understand the Many Worlds interpretation; on the other, you are not sure that you do. You ask yourself "What if the Many Worlds interpretation is inane?" Yes, your inner voice is correct.

The Many Worlds interpretation is not a scientific interpretation. It is not even science fiction, which is supposed to spark imagination and develop creativity. Every science fiction story has at least an ounce of reality.

It is troublesome to see a physics professor in a prestigious university devoting a whole hour of his lecture to explain to his students the Many Words interpretation in earnest, brainwashing young minds with such convolution.

Like in a famous episode of Pauli, which is vividly described by Peter Woit in his book Woit, author of the book *Not Even Wrong: The Failure of String Theory and the Search for Unity in Physical Law*,[31] I say that the Many Worlds interpretation is not even wrong.

The Many Worlds interpretation is a multidimensional runaway absurdity. It is exposed further in other Chapters in this book, including "Flat Earth as Mindset", "The Principle of Proportionality" and Appendix 2.

Super Quantum Mechanics resolves the issue of the Many World interpretation in its totality.

8

Demise of the Schrödinger Cat

In 1935, Schrödinger described his "cat" paradox. The story of Schrödinger's cat has since proliferated scientific literature, lectures and conferences. Any book on quantum mechanics interpretation is considered incomplete unless it includes the Schrödinger cat story. Many thousands of published papers mention Schrödinger's cat in one way or another and decades of endless debates have been devoted to the story. Why then does the story deserve such attention and draw such fascination? Even Schrödinger himself never took his cat story seriously. He considered it ridiculous.

The cat story uses a familiar live macroscopic object to describe a complex microscopic quantum mechanics interpretation. The interpretation says that under certain conditions a cat can be in two quantum states, both dead and alive, with total probability that "dead + alive = 100%."

Although this premise is absurd and has been explained as such in published literature many times, the public remains fascinated by the dead + alive cat: Schrödinger's cat. Even if denounced as an absurdity, such denouncement does not stop the story from being proliferated.

So let us take a closer look. The story goes like this: A hungry cat is put in a closed box. An observer can observe the cat by opening a small window in the box. Behind the box, the same observer places a Geiger counter containing a single unstable nucleus with a decay time of a few hours. The Geiger counter acts as a triggering device. It is attached to an electronics module which in turn is connected to a special driving mechanism. After the unstable nucleus decays, the alpha particle triggers the Geiger counter, and the mechanical device delivers "delicious" food to the cat through the small window in the box. Unfortunately, the food is poisoned. The hungry cat eats the food and dies.

The experiment begins after the Geiger counter has been loaded with the unstable nucleus. The quantum state of the cat "alive" and "dead" is described by wavefunction Ψ_0 (cat) which is a superposition of two quantum states Ψ_1 (alive) and Ψ_2 (dead):

$$\Psi_0 = \Psi_1 + \Psi_2$$

According to one particular interpretation of quantum mechanics, the cat finds itself in two states: dead and alive. Thirty minutes after the experiment begins, you open the window and observe that the cat is still alive. Using your mind you, as an observer, just caused the collapse of Ψ_2 function to zero and reset the timing of unstable nuclear decay.

But why do we mire ourselves in such paradoxes and absurdities that seem almost schizophrenic? We should recall "the image of Einstein standing by his office window, explaining to a visitor that the insane asylum across the way houses those madmen who have not thought about quantum mechanics."[32]

We should recall also Einstein's famous words that quantum mechanics is compelling as a probabilistic science but is incomplete for description of individual quantum entities or events. Attempts to impose probabilistic science on descriptions of individual quantum events or entities cause absurdities, paradoxes, mysteries and enigmas.

If you, a young quantum mechanics physicist, decide to write a doctoral dissertation entitled "Schrödinger's Cat and Some Fundamental Issues of Parallel Universes Branching," good luck to you.

It is not surprising that in his book *A Brief History of Time*, Stephen Hawking exclaimed, "whenever I hear a mention of that cat, I reach for my gun."

Unfortunately, this statement is not sufficient enough to "kill" the Schrödinger cat story. Murray Gell-Mann took a much better approach when he explained in his book *The Quark and the Jaguar*: "The live and dead cat scenarios decohere; there is no interference between them. ...Since the two outcomes decohere, this scenario is not different from a classic one where we open a box inside of which the poor animal, arriving after a long airplane voyage, may be either dead or alive, with some probability for each. Yet, reams of paper have been wasted on the supposedly weird quantum-mechanical state of the cat, both dead and alive at the same time. No real quasi-classical object can exhibit such behavior because interaction with the rest of the universe will lead to decoherence of the alternatives."[33]

I can further expand on the subject. There is no quantum connection between an unstable nucleus and a live cat. There are several classical objects (buffers) between them, such as a Geiger counter, pre-amplifier, amplifier, triggering device and driving mechanism, each of which could fail in a mundane classical way. Poison might not be strong enough, or the cat might be able to detect that there is something wrong with the food. All those silly classical objects would derail the fictitious story of the cat in quantum superposition of live and dead states. And then again, a live cat represents billions of incoherent dynamic quantum states and events occurring in its brain, central nervous system, and living cells. Then, there is also a fundamental question of the cat's consciousness, which is outside of present-day science.

However, in contrast to the Schrödinger cat, there is a viable scenario: two entangled coherent quantum objects. But the Schrödinger cat is not such a case. There is no, and cannot be any, quantum entanglement between an unstable nucleus and a live cat representing a massive set of dynamic incoherent quantum states.

The Schrödinger cat is a pseudo-scientific story and it is a disgrace that some prominent scientists still keep feeding such a story to the general public under the guise of science.

"The final solution" to Schrödinger's cat is the replacement of the quantum mechanics paradigm with the paradigm of Super Quantum Mechanics.

9

Einstein: Supreme Opportunity Lost

Einstein was the most important scientist of the 20th Century. He was also a great philosopher. In the history of science it is rare that a great scientist is also a great philosopher. But what is less recognized is that Einstein was an enthusiastic and talented musician who played the violin and piano since childhood. He once said that had he not been a scientist he would have been a musician. "Life without playing music is inconceivable to me." He played chamber music with his personal friends and often accepted invitations to perform in benefit concerts. Mozart was his favorite composer. He played Mozart beautifully on the violin. This is what Einstein said about the second movement of Mozart's Piano Concerto #21 in C major: "…(it) is like an ideal aria freed of all the limitations of the human voice." Music was Einstein's inspiration when he worked on his physics theories. Einstein was an artist-scientist-philosopher.

As a scientist, Einstein single-handedly opened two scientific curtains: special relativity and general relativity. Both theories are counterintuitive. After several decades, scientists and the general public have accepted both theories without keenly appreciating the theories' counterintuitive paradigm. But what is more relevant to this book is that it was Einstein, together with Max Planck, who began opening the curtain on quantum mechanics. Both were the first contributors to the process of building the foundations of quantum mechanics.

Einstein had an unshakeable commitment to realism. He was disturbed by the huge dose of quantum positivism that he saw coming from the Copenhagen school, especially from Niels Bohr and Heisenberg, during the early years of the development of quantum mechanics. Even today, quantum positivism is entrenched and has proliferated in the interpretation of quantum mechanics. In numerous textbooks and popular articles, you can read such statements as 'a human consciousness causes wavefunction to collapse' or 'an observer creates reality' or 'it is meaningless to talk about the physical state of the system prior to measurement.'

The concept of "measurement" is subjective and confusing. "Measurement" implies "an observer." The concept has proliferated into most physics books and articles. It appears that without "measurements" the universe would not function. What about when the Solar System was in the process of formation? Human observers did not exist. Or when a cosmic gamma-photon hits the surface of Neptune, producing a cascade; that is also a measurement.

In the final analysis, measurement is a physical interaction between two or more quantum entities. Some of the quantum entities might be part of a macroscopic measuring device.

When an individual optical photon reaches the retina of your eye, or a photo detector, or hits the sand in the Sahara desert, it is a remainder of an original gamma-photon produced in the core of the Sun by an individual nuclear fusion event. It takes half a million years and zillions of measurements for an individual gamma-photon—or whatever is left of it—to reach the surface of the Sun and then travel away.

Or take another example of measurement, such as a proton-proton head-to-head collision during the operation of CERN, the Large Hadron Collider. Such a collision produces a jet of secondary particles, which might or might not reach detectors. Detection is actually a secondary issue.

Einstein always stayed firm on the ground of objective reality. He never compromised on the core of his convictions, such as observer-independent objective reality. He never accepted the Copenhagen statements that deemed meaningless the physical state of the system prior to measurement.

Einstein declared, "I am not a positivist. Positivism states that what cannot be observed does not exist. This conception is scientifically indefensible, for it is impossible to make valid affirmations of what people 'can' or 'cannot' observe. One would have to say that 'only what we observe exists' which is obviously false."[34]

Einstein accepted quantum mechanics as a probabilistic theory. But he never accepted quantum mechanics as a complete physics theory. On December 4, 1926, Einstein wrote his famous words: "Quantum mechanics is certainly compelling. But an inner voice tells me that it is not yet the real thing." Einstein also said, "The statistical character of the present theory would then have to be a necessary consequence of the incompleteness of the description of the systems in quantum mechanics..."[35]

Einstein insisted that for a theory to be complete, "every element of physical reality must have a counterpart in the physical theory." He told Heisenberg, "If your theory is right, you will have to tell me sooner or later what the atom *does* when it passes from one stationary state to the next."

Einstein was not bothered with being wrong. At the 1927 Solvay Conference in Brussels, during heated discussions on quantum mechanics foundation issues, Einstein was throwing one gedanken experiment after another at Bohr, keeping Bohr in a state of agitation, out of balance and forcing him to focus his mind.

But Einstein was wrong on one fundamental issue—non-locality. The issue of non-locality is extremely counter-intuitive. His statements about "spooky action at a distance," "ghost waves," "peculiar mechanism," "voodoo forces," "quanta hide" and "telepathic connections" are well known. Einstein was one of the first who noticed non-locality built into quantum mechanics theory. At the 1927 Solvay Conference, in his concluding comments, Einstein gave, as usual, his gedanken experiment as follows: an electron is passing through a pinhole and is detected in a specific location on a semispherical screen; wave function is supposed to arrive at the wide area of the detection screen; once the electron is detected, wave function instantaneously collapses into its location by some "peculiar mechanism" in conflict with his special relativity. *Non-locality was unacceptable to Einstein.*

But in one sense of science history, Einstein was not wrong. The resolution of the non-locality issue was still sometime in the future, beyond his lifetime. During his life, technology was not available to perform experiments of the Clauser/ Aspect kind. We do not know how he would have reacted to John Bell's theorem but he would have certainly given serious consideration to the experimental data.

The separability issue was as important to Einstein as non-locality. Quantum mechanics implies that if two quantum systems interacted in the past and then are separated by any distance, they are entangled. However, Einstein stated, "But on one supposition we should, in my opinion, absolutely hold fast: the real factual situation of the system S2, is independent of what is done with system S1 which is spatially separated from the former."[36]

Einstein was also aware of non-separability of the many particle wavefunctions in configuration (Hilbert) space. He was skeptical of the physical reality of configuration space.

Was Einstein wrong on the non-locality issue? In a narrow sense no, because the resolution of this issue was beyond his lifetime. The curtain on Super Quantum Mechanics was still closed.

On the other hand, by denying non-locality, Einstein deviated from his own philosophy. *He imposed a subjective constraint on objective reality.* However, Einstein proved something important, which is still not recognized by scientists even today. For the last 30 years of his life, he worked on his own in relative isolation from mainstream science attempting to create a deep physics theory beyond probabilistic quantum mechanics. He did not succeed despite his genius and super-human effort. But remember, Einstein is right even when he is wrong.

Einstein "successfully" demonstrated that not even a genius can move the development of quantum mechanics forward to a deeper level if he stays stubbornly within the classical spacetime/energy paradigm and ignores the ontology of non-locality. That was the state of quantum mechanics in the late years of Einstein's life, and that is where quantum mechanics remains to this day—a theory without ontology.

In summary, it is worthwhile to repeat Einstein's philosophical position:

- Science studies objective reality which is independent of any observer.

- Theory must reflect certain aspects of objective reality.

- Quantum positivism is a confused philosophy and an impediment for further progress in quantum mechanics.

- Quantum mechanics is compelling as a probabilistic science but incomplete.

- Elementary quantum entities are real things and must have counterparts in the complete theory.

- Hilbert configuration space is a useful mathematical tool but obscures physical reality.

- Principle of separability is reaffirmed (he was right and wrong on this issue, as discussed in Chapter "Cosmic Seed and Our Universe" in this book).

- And, tragically, he stubbornly rejected non-local dimension.

By rejecting non-local dimension of objective reality, Einstein was unable to make the transition from local "naïve" realism to non-local realism.

Einstein's rejection of a non-local dimension of objective reality was a blunder of monumental proportions. He missed the supreme opportunity of a lifetime to free quantum mechanics from the swamp of quantum positivism and bring quantum mechanics to a deeper level. In his heart, Einstein remained a classical physicist.

For many decades, quantum mechanics has been a hostage of quantum positivism. A shift of the quantum mechanics paradigm is long overdue.

As history shows, great people make great blunders. Einstein remains the most important scientist of 20th Century. He single handedly opened two scientific curtains: special relativity and general relativity, and together with Max Planck was instrumental in opening the curtain on quantum mechanics. This is a reasonable share of scientific contributions for one scientist in his lifetime.

Einstein's philosophy remains a guiding star for scientists. Without a viable philosophy, as without a compass, a theoretical physicist is nothing more than a technician skillful in mathematical formalism.

Part II.

Philosophical Issues

10

"Flat Earth" as Mindset

The Flat Earth concept was dominant for many thousands of years among various human civilizations and cultures, including ancient India, China, early Egypt, Mesopotamia, ancient Near East, and ancient and classic Greece. The great thinkers of those times were as smart and imaginative as the philosophers and thinkers of our time. The Flat Earth model had several variations, such as a flat disk floating in the ocean, a circular earth with a solid roof or a circular disk with an edge, and still other ideas.

As expected, the Flat Earth paradigm had its share of paradoxes. I have mentioned before that paradoxes, enigmas, and mysteries arise from deficient paradigms. For those who subscribe to the Flat Earth paradigm, the well-known paradox would be a sailing ship going into open sea and gradually disappearing from view, not only horizontally but also vertically.

In those times one of the most important questions was, "What's holding up the Earth?" Answers included several concepts, such as that the Earth was supported by pillars, referred to as Pillars of the Heaven; or that the Earth was a cylinder with a flat circular top; and including a ridiculous concept that the Earth was supported by three elephants standing on a giant turtle. Another well-known example, quoted in Stephen Hawking's book *A Brief History of Time*, is of turtles on top of each other, all the way down, holding the Earth.

Among the various concepts of what held up the Earth was one suggested by Anaximenes of Miletus (ancient Greece, 585-528 BCE), who believed that "the earth is flat and rides on air." That was a counterintuitive idea; the Earth is so heavy. How could it hang on air? The concept of Anaximenes of Miletus was prophetic and a breakthrough in human thought ahead of its time.

Why am I writing about such an ancient subject? The reason is that each epoch, including ours, has its own Flat Earth mindset. Even today, with all the progress in astronomy, cosmology and accumulated scientific knowledge, such as general relativity, one still wonders how could such a heavy body as planet Earth hang in space? I hear answers: *It does not. It is being guided by the gravitational force of our sun.* Fine, that is a good answer. But let us expand this subject slightly. We are lucky that our planetary system has only one 'parent' we call Sun. Most planetary systems have two or three or even more 'parents'. Those 'parents' are involved in 'reckless' dynamic behavior forcing their planets (kids) to move along convoluted trajectories. More often than not, these 'parents' lose some of their 'kids.' There are billions of such 'orphans' wandering aimlessly in the Milky Way galaxy.

Now, imagine yourself standing on the surface of such an orphan, a rocky planet the size of Mercury. The sky is filled with stars, nebulae and nearby galaxies. The planet is heavy and hanging in space with no apparent motion. The feeling is eerie. But if this does not impress you, we could use human imagination, which knows no limits, and place the orphan planet in a cosmic void, absent of any stars or even galaxies. The nearest filament of super-clusters is fifty million light years away and can barely be seen. It is really dark, quiet, motionless, as your planet hangs in space seemingly for eternity.

What about the Flat Earth mindset of our time? There are strong indications that current theoretical physics and cosmology are in a crisis as was described by Lee Smolin in his book.[1] Future generations of scientists and philosophers would see our Flat Earth mindset and shake their heads, possibly without realizing that they are in the midst of a Flat Earth mindset of their own time.

A thousand years ago, one perplexing discussion was about Earth; in our time, a perplexing discussion is about multiverse. There are a number of multiverse theories. Most of them are driven by mathematical formalism. However, mathematical formalism is a neutral tool that could bring scientists either to a deeper description of physical reality or to a creation of pseudo-reality.

Nine multiverse theories are listed in the Brian Greene book *The Hidden Reality*[2], such as Quilted, Inflationary, Brane, Cyclic, Landscape, Quantum, Holographic, Simulated, and Ultimate. I am excluding from this list Quantum Multiverse theory, the contemporary name for the Many Worlds Interpretation. The sheer absurdity of Quantum Multiverse theory places it in a class by itself.

None of the multiverse theories succeed in describing our Universe as a member of any multiverse family. The universes produced by these theories have nothing in common with our Universe. All multiverse theories make it appear that our Universe is exclusively unique, which contradicts the General Copernican Principle. Creationists would gladly accept the statement that our Universe is exclusively unique but they would never touch with an eight-foot pole any of the multiverse theories. Creationists have enough troubles with one Universe.

One would think that having one specific universe before their eyes would be helpful to theorists to arrive at a viable multiverse theory where our Universe would be a typical member of a multiverse family. But that is not the case.

Let us challenge theorists to develop a multi-apple theory. We put in front of them an apple bought in a grocery store. Then, when the multi-apple theory is ready, we examine and compare it with a sample apple. Our apple is a masterpiece. It has a unique, individual shape. It has an amazing pattern of colors ranging from red to gold to green. It is fragrant, exquisitely tasty, juicy and crispy. If you cut a small slice and look under the microscope, you will see living vibrant cells. Each cell is a galaxy, full of dynamics, and has incredible fine tuning in all aspects. In contrast, we find that apples described by the multi-apple theory have absolutely nothing in common with the sample apple. All theoretical apples are black and white, have chaotic irregular structures, are sterile, consisting of various mixtures of elements and compounds, and vary in sizes from one micron to one light year. The multi-apple theory produces an infinite number of different types of apples. That is the picture of the current state of multiverse theories.

If you would ask the founders of the multi-apple theory how they would explain the sample apple (from the grocery store), the answer would be 'it exists because it exists.' And they would assure you that you can find any kind of apple out of the infinity of apples with properties arbitrarily defined in advance.

And here is how Brian Greene in his book *The Hidden Reality* describes Inflationary Multiverse: "Eternal cosmological inflation yields an enormous network of bubble universes, of which our universe would be one."[3] One should add that those bubbles are bubbling at an ever-increasing rate from the primordial chaos of eternal chaotic inflation. It is presumed that bubble universes have random and chaotic laws of nature. Obviously, these bubble universes are sterile.

This view is in such stark contrast with our (sample) Universe. Our Universe has immense fine tuning at all levels and in any direction. In Einstein's view our Universe is magical, beautiful and comprehensible.

11

The Principle of Proportionality

The sophistication of a theory is not equivalent to its validity. A theoretical physicist or philosopher must possess intuition and a sense of proportion. Some interpretations of quantum mechanics can be invalidated simply on the basis of their unreasonableness without going into a detailed and time-consuming analysis. To illustrate this, let's start first with some simple conceptual examples.

Suppose you push with your thumb against a wall in your house causing an earthquake in China. Would you accept this cause and effect? The answer is obvious: "No way, it is unreasonable." The result of such action is disproportionate.

Here is another example (and you'll find that my examples keep escalating in absurdity, and still it is impossible for me to get to the level of absurdity published in literature on quantum mechanics interpretations): you are on the equator and you push with all your strength against a wall in a cave toward the west. Do you realize what has happened? You changed the spin of the Earth. From now on, the Sun will rise in the west and set in the east. But you've caused another problem which leads to a catastrophe: now the Earth's spin is opposite to its orbital momentum, slowly but irreversibly causing the Earth to fall into the Sun. This scenario is also completely unreasonable; the effects are totally out of proportion to your action.

(Of course, in accordance with Newton's laws, any push against a wall would transfer no mechanical momentum to Earth, but let us pretend in the name of absurdity that it does.)

A final example: The Andromeda galaxy is 2.5 million light years from the Milky Way galaxy. Both galaxies are heading toward each other with a relative speed of 120 km/sec. Astronomers predict that they will collide in 4 billion years and will eventually merge into one giant elliptical galaxy.

There is a role for you to play. You climb a big mountain, find a giant loose boulder and push it down. The boulder starts falling with an ever-increasing speed, crushing everything in its path. The result of your action is enormous in its consequences: you have caused the Milky Way galaxy to acquire ultra-relativistic velocity in the direction of Andromeda. Now, instead of 4 billion years, it will take only 2.5 million years for both galaxies collide head on. But you know what? Under this scenario, there will be no merger. The Milky Way will pass through Andromeda without even causing a single star-to-star collision. Spiral arm structures of both galaxies would be distorted but the galaxies would not merge. Instead, they will pass through each other. The Milky Way will no longer have a galactic habitable zone. And all of these terrible things were caused by you. Who would believe it? No one would. It is unreasonable and disproportionate.

Let's take another example, no longer related to you, the reader. Imagine that we live three thousand years in the past, in the epoch dominated by the Flat Earth paradigm. We witness an intense, emotional, and profoundly intelligent discussion of the most talented people on the subject of what holds up the Earth. Someone brings forward his theory of an infinite stack of turtles, one on top of the others. This concept elegantly answers the vexing question of what is supporting the next turtle. Is this theory reasonable? The answer is No! And there existed other competing theories just as unreasonable.

Now let us consider de Broglie-Bohm's Pilot Wave – Quantum Potential theory. In this example, I ask you to imagine visiting a quantum research laboratory. With the help of research personnel, you set up a simple experiment. You direct a single electron toward two slits without realizing that, in accordance with de Broglie-Bohm's theory, you have produced an experiment of cosmic proportions. You have created a massive quantum potential field, from cosmic horizon to cosmic horizon, beyond our observable Universe. The quantum potential is guiding the electron travel in a zigzag manner along the narrow groove. Unlike an electrical field which is caused by an electrical charge, or a gravitational field which is caused by gravitational mass, the quantum potential field does not have a source. Furthermore, the quantum potential field has a constant intensity regardless of the distance, even a distance of a hundred billion light years from the laboratory where you performed the experiment. "For example," commented John Bell on Bohm's Quantum Potential theory, "the trajectories that were assigned to the elementary particles were instantaneously changed when anyone moved a magnet anywhere in the universe."[4] The scenario we have here is an extreme case of non-locality. Would you accept such an interpretation? Why not? So many quantum theorists and philosophers spent many years seriously analyzing just such a theory. Some even thought[5] that it was only a historical contingency that had brought the Copenhagen interpretation into the dominant position prior to the arrival of the Quantum Potential interpretation. The Quantum Potential interpretation is outrageously disproportional and unreasonable.

Let us examine the Many Worlds interpretation. Again, you imagine that you are on your way to another quantum optics research laboratory. There, you ask for assistance in performing a simple quantum optics experiment using a single-photon source and a standard beam splitter. Nowadays, single-photon sources are available. You can call them Photon-on-Demand. So you push a button and you get a single photon. You align your source and direct a single photon toward the beam splitter. Do you realize what you have just accomplished according to the Many Worlds interpretation? You have created a parallel universe with a copy of you and copies of all your relatives and friends! Is it reasonable? Many theorists/philosophers consider that it is. They are not surprised that anyone can create parallel universes in his or her spare time. But there is another huge problem with such a theory: individual human consciousness cannot be split into parallel universe branches. Is the belief in such an outcome in proportion with the facts and actions that occurred? I say not.

Let us examine inflation theory. Inflation theory was proposed by Alan Guth and first publicly presented at the Stanford Linear Accelerator Center seminar in January, 1980. The inflation theory has seemingly solved several long-standing cosmological problems. It explained why our Universe has a flat-space geometry and is homogeneous, uniform and isotropic. The concept of inflation is dramatic, but it's also counterintuitive and scientifically, has a "feel good" quality. Then in March of 1980 at another symposium, Guth declared that "inflation, once begun, would continue forever." It is a runaway process with no end. The theory of runaway inflation is in conflict with the Principle of Proportionality. It cannot be a valid theory. The situation is similar to well-known absurdity of "the Ultraviolet Catastrophe" resolved by Max Planck, except here one does not have the luxury of performing an experiment. We need a "Max Planck of our time" to put the theory of inflation on the right track.

Another case to consider is the chaotic eternal inflationary multiverse theory. According to the inflationary multiverse theory, eternal chaotic inflation produces universes ("pocket universes") with random laws of nature at an ever-increasing rate. None of those universes resembles our Universe. The inflationary multiverse theory cannot even explain the only universe that we know and the only universe that is available to us for observation and study, our own Universe. Inflationary multiverse theory thus cannot be a valid theory. It is an extreme materialistic theory and it is in conflict with the Principle of Proportionality.

Let's examine the superstring theory next. The superstring theory is an ambitious attempt to explain all of the particles and fundamental forces in our Universe. In addition to three spatial dimensions, it adds seven more spatial "hidden dimensions." The superstring theory produces 10^{500} different universes, each with its own random laws of nature. " To get an idea how many that is, think about this: If some being could analyze the laws predicted for each of those universes in just one millisecond and had started working on it at the big bang, at present that being would have studied just 10^{20} of them.

And that's without coffee breaks," say Stephen Hawking and Leonard Mlodinov in their book, *The Grand Design.*[6]

The superstring theory is not valid because it is in conflict with the Principle of Proportionality. Fundamental reasons why the superstring theory is not valid are given later in this book.

One could argue that the Principle of Proportionality is not a scientific method. Perhaps not, but once applied it would save many years in the professional life of a scientist/philosopher that would otherwise be wasted on non-productive, meaningless activities based on absurd theories and interpretations.

12

Humans as an Embryonic-Stage Intelligence

We, humans, are at an embryonic intelligence stage of development. But we are presumptuous. We have a high opinion of ourselves and of our place in the Universe. Agreed, potentially, we have a great future. But we are not there yet. As a civilization, we are only ten thousand years old. Our science is only a few hundred years of age. Even dinosaurs had 180 million years before their demise. It is difficult for us to imagine where we will be one thousand or ten thousand years from now, or one hundred thousand, or one million years, or one hundred million years from now. It is mind-boggling to even think about it. But what is ten thousand years compared to the age of our Universe, 14 billion years? On a cosmic scale, our ten thousand years is a mere nanosecond.

Our Universe is a middle-aged entity; it has possibly another 15 billion years to go before its cycle is complete and the Universe folds out of existence. How will the end occur? Science does not know yet.

What is our place in the Universe? Our planet is a dot in the Solar system. Our Solar system is a dot in our Milky Way galaxy. Our galaxy is a dot in the observable part of our Universe. The observable part of our Universe is a dot in our Universe. And our Universe is one out of trillions of other universes in Uni-Universe (see Chapter "Our Universe and Uni-Universe").

Some might be offended by the term embryonic-stage intelligence. But in fact this term is constructive and optimistic. It shows what a great potential future we, humans, should expect.

In summary, here are the parameters of human existence: embryonic-stage intelligence, one nanosecond, and a dot.

But we, humans, are on the path of evolution toward an advanced intelligence stage, barring a potential catastrophe either natural or of our making.

Please refer to Appendix 3: Chicks Story.

13

The Law of Fine Tuning

In this Chapter I am replacing the Anthropic Principle, which is archaic and overused, by the Law of Fine Tuning.

If you place on the table all physical constants such as gravitational constant, Planck constant, velocity of light, cosmological constant, electric charge, and also the particular properties of the basic building blocks, such as electron, neutrino, quark, gluon, and so on, you would have an incredible collection in front of you. Now, try to make sense of this huge puzzle and find interrelations among different pieces. The impression you would get is that all those physical constants and properties are totally random. But any attempt to change the value of any piece of the puzzle, however slightly, would transform our Universe from habitable to sterile.

I make a heuristic assumption: *Any change, however slight, in the value of any physical constant cannot be compensated by any adjustments in any other physical constant or constants.* The previously-mentioned collection of constants and physical properties is unique and represents *a sharp definition* of our Universe. The laws of nature, including the laws of physics with their constants, are preset and fine-tuned to "absolute" perfection prior to the Big Bang.

Some scientists are amazed that Earth's orbit is in a precise location for life to evolve. One can explain this fact, as many astronomers do, as mere coincidence. One can explain away one, ten or even twenty coincidences, but after you reach eighty coincidences, the odds of that happening are 1 out of 10^{80}. To understand how small these odds are, imagine that all protons in our observable Universe are blue with the exception of one, which is red. Then, with your eyes closed and in the midst of these countless protons, you blindly stretch out your hand and randomly pick one proton. Voila! It is red! What a coincidence!

As I mentioned previously, Earth's orbit is in a precise location. The location is not only precise but also stable from the time of the formation of the Solar system. This is an important requirement for life to evolve on planet Earth. In fact, stability and integrity of the Solar system is the strict requirement for evolution of life on Earth. Each planet of the Solar system has a precise and stable orbit. Each planet plays a supporting role in evolution of life on Earth. Some materialistic theories claim that after the formation of the Solar system some planets had wandered wildly in the planetary space causing the late heavy bombardment.[7] These theories must be re-examined. An alternate explanation of the late heavy bombardment is offered in Reference 2 of the Introduction.

There exist two categories of fine tuning: *universal* and *local*. The universal fine tuning presets the laws of nature including the laws of physics, to perfection, prior to the Big Bang. It assures dynamics of physical processes in our Universe on *autopilot*. The universal fine tuning is necessary but not sufficient to create physical conditions for origin of life in our Universe. The universal fine tuning is a justification for materialism (to read more about this, see Conclusion).

The local fine tuning brings all required conditions and environment for origin and evolution of life in *qualified locations*. There are very few qualified locations. The rest of the universe operates on autopilot and is not suitable for life. Our Universe model is inefficient.

The local fine tuning takes place after the Big Bang. Results of the local fine tuning can be found in all qualified locations and at all levels in our Universe, including galaxies with galactic habitable zones, habitable planetary systems, habitable planets, all forms of life including humans, living cells, cellular components such as protein and DNA, and quantum entities. Every living cell is like a functioning galaxy with its metabolism superbly synchronized and fine-tuned. The local fine tuning is a creative force, responsible for the origin and evolution of life.

The existence of such a massive number of universal and local fine-tuning factors is a fact of objective reality whether one likes it or not. We should not bury our heads in the proverbial sand.

Some scientists have a tendency to evoke the name of God. The concept of God is beyond human comprehension and is not helpful to science. At this time, science does not have explanations as to why our Universe has such a massive number of fine-tuning factors. We just do not know. But there is no reason to deny the fact. As an embryonic-stage intelligence, humans have little knowledge about objective reality. Many scientific challenges still lie ahead of us.

14

Science and Human Visualization

The Law of Fine Tuning postulates the existence of only three spatial dimensions (3D) on all levels, starting from our Universe itself as the entity, down to galaxies, stars, planetary systems, all rock formations, all forms of life, subcomponents of life forms such as DNA and protein, and all basic building-block particles, such as photons, electrons, neutrinos, quarks and gluons. All undiscovered deeper levels of objective reality in our Universe are also three-dimensional, and there are no exceptions. The human brain is three-dimensional. Human imagination is also three-dimensional. Nature does not suddenly resort to more spatial dimensions at some certain deep level. That does not happen.

Niels Bohr underestimated the human ability to visualize *a quantum world* which is also *three-dimensional*. He stated, "...we must be prepared for the necessity of an ever extending abstraction from our customary demands for a directly visualizable description of nature."[8]

Progress in natural sciences is impossible without human visualization. With no progress in natural sciences, evolution of humans from our present embryonic-stage intelligence toward the advanced level of intelligence is not possible. But evolution is a law of nature supported by the Law of Fine Tuning.

Humans cannot visualize images which are more than 3D. Try to visualize a simple object such as a 4D cube.

The objective reality outside of our Universe may have more than three spatial dimensions but we, humans, are locked into our Universe's 3D reality.

The superstring theory with seven hidden spatial dimensions describes a pseudo-reality.

15

"End of Physics" as Myth

Many times physicists have pronounced that physics is completed and there is not much left to do. "The more important fundamental laws and facts of physical science have all been discovered.... Our future discoveries must be looked for in the sixth place of decimals," stated Albert A. Michelson in his speech at the dedication of Ryerson Physics Lab, University of Chicago over a hundred years ago in 1894.

In 1900, Lord Kelvin (William Thomson), an influential British scientist gave a lecture to the Royal Institution of Great Britain, where he famously proclaimed: "There is nothing new to be discovered in physics now. Our work has been done. All that remain are two tiny clouds on the horizon."

I do not have Lord Kelvin's famous lecture handy, therefore I will use my imagination to reconstruct it:

"Ladies and Gentlemen, we are here to congratulate ourselves. The body of physics is finally completed. Such monumental work has been accomplished by the dedicated effort of many scientists throughout several centuries beginning with Galileo and continued by Newton, Kepler, Dalton, Lavoisier, Faraday, Maxwell, Clausius, Poincare, Lorentz, and many other scientists including your humble servant. On the blue sky of physics, only two tiny clouds remain. One is the Michelson-Mosley experiment in measurement of Earth's velocity relative to ether. The experiment conflicts with the existence of ether. One thing we are confident of is the reality and substantiality of the luminiferous ether. One should re-examine what is wrong with the experiment because none of us doubt the existence of ether. The second cloud pertains to a runaway problem in the theoretical explanation of the black body radiation spectrum in

the ultraviolet range. We should assist those theorists in solving the black body radiation problem.

By the way, my nephew Peter Thomson, will be graduating this year from Cambridge University with a Bachelor's degree in Natural Sciences. He wants to do physics. I have advised him against it. Not much is left to do. All that remains is more and more precise measurement. A generation of young scientists has missed the opportunity to work on physics problems. I have recommended him to take a closer look at botany. There are ongoing amazing discoveries of new species in the Amazon Rain Forest basin."

What actually happened is now a well-known chapter of physics history. Both of the tiny clouds rapidly grew into giant thunderstorms, transforming the whole landscape of physics beyond recognition. From one cloud, quantum mechanics was born by the efforts of Max Planck, in 1900 (the same year of Lord Kelvin's speech), and Albert Einstein in 1905. The second cloud, through the efforts of Einstein, also in 1905, grew into the special theory of relativity, and in 1916 into the general theory of relativity.

This is a dramatic lesson in science that we all must heed.

We hear over and over again about "ultimate theory", "unified theory" and "theory of everything." In his book *The Theory of Everything*, Stephen Hawking predicted "...I still believe there are grounds for cautious optimism that we may now be near the end of the search for the ultimate laws of nature."[9] Later, he changed his mind.

The more scientists talk about the end of physics, the more likely a fundamental scientific breakthrough is about to occur. A new curtain would open and behind it, one would find an unexpected element of reality so counterintuitive that it could not be predicted either by logic alone or by sophisticated mathematical formalism. As an embryonic-stage intelligence, we humans should expect a countless number of such curtains ahead of us. Physics and science in general are open-ended and progress is never ending.

The next curtain is in the process of opening. It pertains to non-local realism and Super Quantum Mechanics.

16

Science and Curtains

As humans, we are in the early stages of evolution toward an advanced intelligence level. Our perception of reality is limited. There is much more in existence than we know. Science opens one curtain after another. Behind these curtains is an amazing world of reality, totally unexpected and counterintuitive. Each curtain that opens is a fundamental scientific breakthrough and a new paradigm. When scientists are facing a new curtain, logic alone or a sophisticated mathematical formalism is not sufficient. It requires intuition and counter-intuition.

The history of science gives many examples of such curtains in physics, cosmology, biology and other branches of science. Here are a few examples.

Copernicus Theory

For fourteen hundred years, until 1543, Western civilization was dominated by the Ptolemy/Aristotle model of the cosmos. Since the time of Aristotle, it was believed that Earth was the stationary center of the Universe, and in Ptolemy's model, Earth stood still at the center, and the Sun, planets and stars moved in complicated orbits described by epicycles. The Ptolemy model was readily adopted by the Catholic Church, held as official doctrine and enforced by the Inquisition. The Ptolemy/Aristotle model was an impediment to intellectual progress in Europe for many centuries. Then, in 1543, Copernicus published his book *De revolutionibus orbium coelestium (On the Revolutions of the Celestial Spheres)*. He worked on his theory for several decades and published it just prior to his death. His theory describes that Earth is not the center of the world but is one of six planets revolving around the Sun. Such an incredible idea met strong resistance and led to passionate debates. It was considered contrary to the Holy Scripture and Copernicus's followers, including Galileo, were tried for heresy.

What is important to us is the paradigmatic aspect of the Copernican concept. The concept was counterintuitive. It was difficult to believe that Earth was moving in a circular orbit through the cosmos, and people and everything else were traveling along. But the Copernican concept was much more than about the position of planet Earth. If not Earth or Sun, then what is the center of the world? General Copernican principle dictates that there is no center. None of the existing entities in the cosmos can claim to be the center. In fact, Copernicus himself would not have likely imagined that his concept was of such transcendental significance as to be transformed into the General Copernican principle or the GC principle, which I will expand in the companion book (see Reference 1 of the Introduction). The GC principle's full impact is yet to be understood.

The GC principle is formulated as follows: *In objective reality no one entity can claim an exclusive uniqueness either in space, time, or in substance.* The GC principle has tremendous predictive power.

As a habitable planet, Earth cannot be the sole habitable planet in our Universe. As a galaxy with a galactic habitable zone, the Milky Way galaxy cannot be the only habitable galaxy in our Universe. Our Universe has to be finite in size and cannot be exclusively unique. As the habitable Universe, it cannot be the only one in the Uni-Universe.

Electromagnetism and Maxwell Theory

In the 18-19th Centuries, physicists in several countries conducted experimental studies of electricity and magnetism. Gradually, especially through the work of Danish physicist Hans Cristian Øerstead and British scientist Michael Faraday, it was discovered that electricity and magnetism are interrelated. One can produce magnetic force by moving electric charges or produce electricity by moving magnets. But the greatest mystery was how those forces act through the empty space separating the interacting objects. The predominant mindset was that in order to move an object something else must come in contact with it, pulling or pushing. Faraday proposed a concept of invisible tubes in space between magnets or electric charges. He assumed that those tubes were the ones which provide mechanical interactive force.

It was the Scottish physicist, James Clerk Maxwell, who in 1864 developed the classical electromagnetic theory which united electric and magnetic forces and offered an explanation of the electromagnetic nature of light. The principal idea was that an electromagnetic field propagates through space as a wave and produces action on objects. It was a giant scientific step forward.

With further progress in science, the concept of field acquired key significance. Eventually, a carrier of electromagnetic force was established—photon—explaining that an exchange of photons between two electrons produces electron-electron interaction. This concept was extended to other forces, such as weak nuclear force, strong nuclear force, and gravity. It was established that each category of field had its carrier or carriers. The concept of field has proliferated into quantum physics in the form of quantum electrodynamics (QED) and quantum chromodynamics (QCD). The curtain was now wide open. From the simple phenomena of electricity and magnetism, a new vast aspect of objective reality was unveiled. The counterintuitive concept of field acting in empty space became an accepted view.

Special Relativity

Einstein opened the curtain on special relativity in 1905. Special relativity includes counterintuitive features such as length contraction, time dilation and relativity of simultaneity. The theory predicts the equivalent of matter and energy as expressed in mass-energy equivalence formula $E = mc^2$. The theory predicts that c is not just the velocity of electromagnetic wave in free space, but a universal constant of spacetime. The theory says that it is impossible for any particle with rest mass to be accelerated to the speed of light. The theory incorporates the principle that the speed of light is the same in all inertial frames of reference.

Today, every student takes special relativity for granted. It is a part of us and we are not surprised by its relativistic effects. But there was a time, before 1905, the year that Einstein discovered special relativity, when such a concept would have been mind boggling.

The following is a well-known twin paradox. Suppose we are living before Einstein's discovery of special relativity. Imagine two twin brothers in their 20s. One brother leaves in a super spaceship on a long-range space travel, reaching ultra-relativistic velocity close to the velocity of light. The other twin remains on Earth and continues his everyday life. After one year of his time, the first twin, still in his 20s, returns to Earth and discovers that his brother is very old, in his 80s. Before the discovery of special relativity, the idea would have boggled the mind. But today? It is not so surprising any more. High energy physicists see such phenomena in their experiments every day: the life of unstable particles is much longer when they travel with ultra-relativistic velocity.

Of course, the story of the twins is conceptual. Space travels beyond our Solar system are not feasible. Planet Earth is friendly to humans while our Universe is hostile. For example, it is not feasible for humans to travel throughout our Milky Way galaxy. Long-range space travel away from our home planetary system is *a travel into obsolescence*. For humans, long-range space travel in a *physical* form is not feasible regardless of the state of technology.

General Relativity

General relativity or the general theory of relativity was developed and published by Albert Einstein in 1916. The substance of the theory was a description of gravity as a geometric property of space and time, or spacetime. The general relativity predictions have been confirmed in observations and experiments.

Scientists such as Schwarzschild, Friedmann, and Lemaître, among others, developed further various aspects of the theory.

The theory has important cosmological implications. A combination of theory, observations and experiments opened a new curtain in science, offering counterintuitive properties and features of a cosmological world where gravity is a principal factor. Rather than being eternal and static, as Einstein thought, the Universe is expanding with retroactive projection to its origin in the form of the Big Bang 14 billion years in the past.

The following counterintuitive properties and features have been theoretically predicted, cosmologically observed and/ or experimentally confirmed: gravitational time dilation, the gravitational frequency shift such as redshift or blueshift of light, gravitational time delay, bending of light, gravitational lensing, gravitational waves, and the existence of black holes.

Einstein's equations are non-linear. Matter objects define spacetime geometry and, in turn, spacetime geometry defines matter objects' motion. One of the extreme examples is a merger of two black holes in a wild cosmic dance.

Special relativity is a subset of general relativity where the factor of gravity is small.

Quantum Mechanics

As previously described, in 1900 Max Planck and in 1905 Einstein made a counterintuitive breakthrough and opened the curtain on a strange and mysterious microscopic quantum world. During the 1920-30s, the theory of quantum mechanics was developed principally through the efforts of Bohr, Heisenberg, Schrödinger, Max Born, Dirac, Pauli and von Neumann. So far, quantum mechanics has never failed any statistical predictions.

Einstein kept insisting that quantum mechanics is compelling as a statistical science but is incomplete as a theory because it does not include a description and explanation of individual quantum events. Since the 1930s, quantum mechanics has grown into quantum electrodynamics and quantum chromodynamics, providing an important tool for studying the properties of basic particles. However, in spite of its many interpretations, quantum mechanics remains an enigma that as yet, explains nothing.

Transition to Super Quantum Mechanics

Scientists have noted a mysterious non-locality from the beginning foundations of quantum mechanics. Einstein stubbornly refused to accept a non-local reality. In 1964 John Bell developed his theorem addressing the non-locality issue. Several experiments were performed, especially by Stuart Freedman and John Clauser (1972) and Alain Aspect with his collaborators (1982), which demonstrated the existence of non-local influences. John Bell's scientific discovery and follow up experiments were the beginning of an important transition from quantum mechanics to Super Quantum Mechanics.

Super Quantum Mechanics

With Super Quantum Mechanics a new curtain into the quantum world is about to be opened. It is counterintuitive. In addition to the physical dimension, Super Quantum Mechanics incorporates a non-locality dimension. Super Quantum Mechanics manifests that the era of quantum positivism is over. Quantum entities with all their attributes are real and not hanging in limbo, whether their pre-existing conditions prior to measurement are known or not. Instantaneous non-local interactions among individual quantum entities are real, with no conflict with special relativity.

Super Quantum Mechanics brings ontology to quantum mechanics. A new paradigm explains all mysteries, enigmas and paradoxes.

Part III.

Current Cosmology and Physics Issues

17

Theory of Inflation

The founder of the inflation theory is Alan Guth, Ph.D. from MIT. He said in his book *The Inflationary Universe*, "I view the official debut of inflation as the seminar that I gave at SLAC on January 23, 1980."[1] That seminar was the first time the scientific world heard about the inflation theory.

The inflation theory is an attempt to explain why the Universe is flat, homogeneous, uniform and isotropic. Many cosmologists and theoretical physicists assume that at the very beginning, the 'starting state' of the Universe prior to the Big Bang, the Universe was violent, turbulent, with a highly non-uniform searing temperature, high gradient density distribution, and convoluted space geometry. This is how it is stated by Paul J. Steinhard and Neil Turok, "Cosmologists differ on the precise properties of this starting state, but many believe it would have been wildly turbulent and non-uniform, with huge variations in density and temperature from place to place..."[2] That is a materialist point of view. And how did it happen that presently the Universe is flat, uniform, isotropic and homogeneous? The inflation theory explains that for an extremely short period of time, such as 10^{-30} seconds, the Universe grew explosively, doubling its size every 10^{-35} seconds. And, in the period of 10^{-30}, the Universe doubled in size 100,000 times, solving the cosmological problems: flatness, uniformity, isotropy and homogeneity. The theory also explains why the Universe is so large.

Later that year at a seminar at Harvard University in March of 1980, Guth stated that the inflation theory explained several long-standing cosmological problems. That was the good news. Then he dropped the bomb: "The very mechanism that solved the cosmological problems made it impossible for the rapid expansion to end. Inflation, once begun, would continue forever."[3]

The concept of inflation is dramatic and counterintuitive. It seemed reasonable until the founder of the inflation theory stated that inflation, once begun, would never stop. Has any scientist ever encountered a natural runaway process that never ends? The answer is no. Consider such natural processes as a snow avalanche, nuclear chain reaction, explosion of supernova, or the collapse of a giant star into a black hole... all occurrences in nature have a beginning and an end.

The theory of runaway inflation cannot be a valid theory. Assuming that inflation is a real natural phenomenon, it must have a beginning and an end. The inflation must cease orderly, in contrast to eternal inflation. A challenge for Guth and other scientists is to develop a theory which would reflect the orderly completion of inflation. The theory of inflation would be considered a viable theory if it could show that inflation stops in an orderly manner.

Alan Guth claims that a huge expanded universe was produced from almost nothing. He said, "in the context of inflationary cosmology, it is fair to say that the universe is the ultimate free lunch." This statement is bravado and absurd. One cannot produce a viable universe using only space geometry, energy density, temperature and time; that is too simplistic.

In the Chapter "Cosmic Seed and the Origin of Our Universe" I offer an alternative concept of the origin of the universe with an understanding that the full scientific explanation of the origin is not achievable by embryonic-stage intelligence. This statement should not be taken as a discouragement. It is an indication again and again that there is no end to scientific progress.

18

Inflationary Multiverse Theory

The principal author of Inflationary Multiverse theory is Andrei Linde.[4] The theory is a bizarre and grotesque description of physical reality. The driving forces in multiverse are primordial chaos, eternal chaotic inflation, and quantum fluctuations. All these forces keep producing random universes ("pocket universes") with random laws of nature at an ever-increasing rate including the production of pocket universes by other pocket universes. It is a bizarre theory resulting in a runaway process of ongoing production of pocket universes without beginning and without end. There is a huge mismatch; the typical time required for the production of a pocket universe is 10^{-30} seconds and the typical lifetime of a universe is estimated as 10^{11} years. The inflationary multiverse theory describes an incredible situation: trillions and trillions of pocket universes are being pumped out at an ever-increasing rate into objective reality with a very restricted exit. It is akin to some kind of cosmological "constipation." Moreover, universes in the multiverse theory do not resemble our Universe, which has a certain order, structure, massive fine tuning, harmony, majesty and breathtaking beauty.

We know from studying and observing our Universe what incredible fine tuning is required to produce a universe suitable for life and the evolution of intelligence. Fine tuning is everywhere and at all levels in our Universe. How does the multiverse theory explain that the only Universe available to us for observation and study fails to fit into the multiverse description? Authors of multiverse theory explain it by invoking the anthropic principle, which states (in my interpretation) that out of an infinite number of possible random universes, one can always find a universe with properties arbitrarily assigned in advance.

Multiverse is a materialist theory. Materialism is not compatible with the anthropic principle. It is pathetic for materialistic scientists to appeal to the anthropic principle.

Multiverse theory also conflicts with the General Copernican principle which states that no one entity may claim an exclusively unique position in objective reality. Multiverse theory assigns to our Universe an exclusive and privileged status: we are a kind of exclusively unique center of objective reality and our Universe is a miracle. On this point, creationists readily agree. They say, "Yes, indeed, it is miracle. We told you so. God created the world for the benefit of humans." As one can see, materialists and creationists share the same chair. Two opposites meet.

We all know the statement by Einstein that "it is incomprehensible that the Universe is comprehensible." With respect to multiverse theory, one should paraphrase Einstein in reverse: "It is comprehensible that the inflationary multiverse theory is incomprehensible."

19

Superstrings or Not Superstrings?

For the last three decades, a group of super-talented theoretical physicists has been working on the superstring theory with an ambitious goal of explaining all of the particles and fundamental forces in our Universe. It is an attempt to arrive at the ultimate physical theory. As previously stated in this book, the 'ultimate' is never ultimate. But this is not the principal point. Actually, one does not even need to know the details of the superstring theory to issue a verdict: the superstring theory is not a valid theory.

This book is not a review of the superstring theory. You can find a detailed analysis of the theory from two of its vocal critics: Lee Smolin who wrote the book *The Trouble with Physics: the Rise of String Theory, the Fall of a Science, and What Comes Next*[5], and Peter Woit, author of the book *Not Even Wrong: The Failure of String Theory and the Search for Unity in Physical Law.*[6]

In my view, there are at least three fundamental reasons why the superstring theory is an immensely misguided effort leading to a Never Never Land.

Reason one is philosophical. The superstring theory is one more example where a particular mindset—not lack of super-scientific talent—drives development of a physics theory towards an impasse.

Reasons two and three are foundational. There exists another aspect of objective reality of which theoretical physicists in general and superstring theorists in particular are oblivious.

For these three reasons, specifics of the superstring theory are irrelevant.

Reason One (Philosophical)

The superstring theory assumes the existence of three actual space dimensions and seven "hidden" ones. But 'hidden spatial dimensions' is a misconception. Such hidden dimensions do not exist. They are in conflict with the Law of Fine Tuning, which postulates the existence of only three spatial dimensions (3D) in our Universe at all levels.

As discussed in the Chapter "Science and Visualization," humans can visualize only in three-dimensional space. Without visualization, progress in natural sciences is impossible. Without progress in natural sciences, the evolution of humans from our current, embryonic-stage of intelligence toward advanced intelligence is not possible. But evolution is a law of nature supported by massive fine tuning.

The development of the superstring theory is driven by sophisticated mathematical formalism, resulting in grotesque abstract structures of D-branes of various orders, with no apparent relation to objective reality. The physical theory is gradually being transformed into a complex mathematical edifice. The theory has produced an infinite number of solutions but made not a single testable prediction. Other scientists and the general public have greater and greater difficulty following the progress of the superstring theory. Superstring theorists are holding the tail of mathematical formalism in the hope that it would take them somewhere. But mathematics is neutral. If the philosophical and foundational concepts are wrong, the mathematics would describe pseudo-reality.

Reason Two (Foundational)

Superstring theorists have overlooked non-locality dimension, an important aspect of objective reality. Future physics theories should include both dimensions: physical and non-locality (not to be confused with spatial dimensions!). One day the superstring theorists will realize this and exclaim 'Oops, out of two dimensions we have overlooked one!'

Reason Three (Foundational)

According to Super Quantum Mechanics, the principal mode of existence of elementary particles, both matter and force, is spin. That includes spin-0 particles, which is counter-intuitive but true. The elementary particles are spinning and spinning. Humans can easily visualize such spinning in 3D space.

Contrary to the Standard Model, elementary particles are not pointlike with no structure. Neither are they extremely small vibrating energy strings as is assumed in the super-string theory.

The elementary particles have structure and finite size.

For these three reasons, the model of vibrating energy strings is a misconception.

Part IV.

Advanced Ideas in Cosmology

20

Our Universe and Uni-Universe

I am introducing the term Uni-Universe as an alternative to multiverse. There exist several multiverse theories, as presented in Brian Greene's *Hidden Reality*[1]. Each theory defines multiverse as an infinite number of universes, each with random laws of nature. By this definition and in accordance with the Law of Fine Tuning, each of these universes is sterile and has nothing in common with our Universe, which is the only example of a universe that we know. In contrast, Uni-Universe has billions or even trillions of universes where each individual universe is habitable and suitable for the evolution of life and intelligence toward an advanced stage.

Uni-Universe exists in a four-dimensional space, which cannot be visualized by humans. Individual universes are like three-dimensional entities floating in the four-dimensional space of Uni-Universe. Each individual universe has a sharp definition. It is uniquely and sharply defined by a combination of fundamental constants, fundamental properties and the laws of nature. No two universes are identical.

The odds are overwhelming that our Universe is average and not exceptional. Our Universe is finite in size and governed everywhere by the same laws of nature. Suppose we travel instantaneously in our imagination, from the observable part of our Universe to some distant part located, let us say, one hundred trillion light years away, bringing with us all of our cosmological instrumentation. We would find no detectable difference.

Our Universe is impregnated with countless "life dots"—habitable planetary systems. There exists only one principal form of life throughout our Universe, which is specific to our

Universe. It is carbon-based. The other rudimentary forms of life on the fringes have no potential. There is no strict synchronization of life evolution among habitable planets. Each individual evolution cycle has a different profile, speed and timing. Habitable planetary systems are separated by vast distances to prevent disturbances and influences of one system on another. This is another example of local fine tuning.

On some planetary systems the evolution of life from its origin to the embryonic-stage intelligence takes only one billion years or even less, whereas on Earth it took 3.8 billion years. It appears that far in the past, before the Cambrian explosion, the Solar system ran into a huge but anticipated calamity, such as a giant cosmic dust cloud—remnants of a supernova. The solar system and the cosmic dust cloud might have drifted together for an extended period of time along a galactic orbit. This had delayed the evolution of life on Earth for as much as 2 to 3 billion years. It is not by chance that the primordial Sun, after its formation, shed 4% to 7% of its mass[2] thus extending and adjusting its life by approximately 3 billion years.

As the Cambrian explosion shows, life evolution can be very fast. Or take the example of the evolution of mammals after the demise of dinosaurs: in a short 65 million years, a small, insignificant creature which hid underground evolved into man.

21

How Large is Our Universe?

Science assumes that just prior to the Big Bang our Universe was the size of an atom or even smaller. How large is our Universe at the present time? It's a good question. Some cosmologists say 'the Universe is infinite.' I say not likely. Nothing is infinite.

If objective reality consists of nothing but the infinite Universe, then it would be "infinite sameness." Objective reality does not engender sameness. Objective reality never tires of bringing counterintuitive surprises.

But still, what about the size of our Universe?

There was a time in European history when the burning question was "what is the size of Earth?" At that time, the Ptolemy/Aristotle model was the dominant European worldview, embraced by the Catholic Church and enforced by the Inquisition. As a result, European intellectual progress was impeded for many centuries. A Copernican revolution was overdue and should have happened several centuries earlier. At that time, people did not know how large Earth was although many pointers were staring them in the face.

Imagine that you are living during that epoch. To answer the question about the size of Earth you would not need advanced technology or any technology. The only thing you need to know is how to measure length. For example, you could select a place on a coast, opposite from a visible island. Then, on the beach at sea level, install a tall post with clear markings. Looking from your island position on a clear day when the sea is quiet, you would observe what portion of the post went down and could no longer be seen. Now you have to measure the distance from your observation position to the post.

Perhaps, it is not easy but with some effort, it is doable. Remember, you are attempting to estimate the size of the Earth! Once it is done, it will be a huge accomplishment!

Alternatively, you may want to use a sailing ship or a boat and install a really giant post on a coast that can be seen from a long distance. After all, building a really giant post is still not as challenging as building an Egyptian pyramid. But such a thought did not occur to people of those times.

Of course, if you had portable clocks, then in addition to length measurement, you can measure time. You and a friend could travel to the equator during a suitable season when the sun is straight up at its zenith, and proceed with your measurements. You have to synchronize your clocks and start walking away from each other in opposite directions along the equator in order to be separated by as large a distance as practically possible. Then each of you independently record the time at which the sun is exactly at zenith and no shadows are seen. The time difference and the distance give you an estimated size of Earth.

During those times there were other options for the estimating the size of the Earth, including Moon and Sun eclipses. But people were oblivious. The size of Earth was simply staring at them, waiting to be discovered.

A similar situation exists now. We are anxious to find the size of our Universe. Obviously, 'an infinite size' would not be a satisfactory answer. It would be a short cut. If you want to impress someone, say 'infinite.' But remember, nothing is infinite. Infinity is a mathematical concept, a useful tool.

The progress in cosmology and science in general is being impeded by our inability to estimate the size of our Universe. This inability also breeds various dubious multiverse theories. Like many hundreds of years ago with respect to the size of Earth, the size of our Universe is staring us right in the face but we are totally oblivious.

It is unlikely that if we would look through the most powerful telescope or use the most sophisticated cosmological instrumentation, we would find the answer. Eventually, nothing can prevent humans from finding the answer. It will be totally unexpected and mind-boggling. It is an excellent problem for cosmologists and theoretical physicists to ponder. Objective reality is keeping in its store a counterintuitive answer.

In the next Chapter, "Application of the Law of Fine-Tuning in Cosmology: Study of Three Cases," I attempt to estimate the size of our Universe.

22

Application of the Law of Fine Tuning in Cosmology:

Study of Three Cases

Our Universe is fine-tuned including pre-Big Bang universal fine tuning and post-Big Bang local fine tuning. Scientists have discovered hundreds and hundreds of fine tuning factors.[3] There is no way one can explain such fine tuning by mere coincidences. Each fine tuning factor is multi-purpose and covers several seemingly unrelated areas. Local fine tuning is required to create a habitable cosmic environment, such as habitable galaxies with their galactic habitable zones, habitable planetary systems, and habitable planets.

Our Universe is sharply defined in terms of its precise and "absolute" values of physical constants and the laws of nature. A small change in the value of any one constant leads to the transformation of our Universe from being capable of supporting life to one which is sterile. A small change in the value of any one physical constant cannot be compensated by changes in any other constants. All these observations lead to the conclusion that intelligence does not reside in humans only. Nature is also intelligent, and consciousness is a property of objective reality. However, the existing scientific paradigms are not adequate to address and explain the issue of consciousness. Perhaps we humans have not yet reached the stage in our evolution to understand consciousness. At a minimum, we have to recognize our limitations in this area.

The Law of Fine Tuning has tremendous predictive power. For example, there is a well-known story about the pre-eminent astronomer Fred Hoyle who predicted fine tuning in the rate of production of carbon and nitrogen in the hot interiors of red giant stars. However, the Law of Fine Tuning is not a substitute for existing scientific methods; but it can be used as a complementary tool with great effect.

In the following *original* study, I will demonstrate the application of the Law of Fine Tuning in three cosmological cases.

Case One – Perfect Solar Eclipses

We live in an epoch of perfect Solar eclipses, also called total eclipses. Perfect Solar eclipses are starkly different from partial or annual eclipses.

After the formation of the Earth-Moon system 4.5 billion years ago, the Moon was six times closer to Earth than it is now. Approximately four billion years ago Earth was totally covered by water and was spinning four times faster than it is today, causing strong winds. Being close to Earth, the Moon exerted a strong gravitational tidal force producing enormous tidal waves. At that time, a projected image of the Moon would have far exceeded a projected image of the Sun. No perfect Solar eclipses existed. As a result of a great expenditure of gravitational tidal energy, the Moon kept receding from Earth and slowing down Earth's rotation. In fact, it was receding much faster than today.

If we look in the distant future, say 4 billion years from now, the Sun will begin entering the red giant stage, growing in size, and the Moon will move farther away from the Earth. A projected image of the Sun will substantially exceed the projected image of the Moon. Again, no perfect Solar eclipses will be possible. It is another "strange coincidence" that the sizes of both projected images of the Sun and the Moon were so close to each other during the epoch when humans arrived on the scene. How is it possible? We find different and somewhat conflicting explanations.

Some scientists say that it is just a coincidence or one of those strange coincidences. Creationists say, "No. It is not a coincidence. Perfect Solar eclipses are spiritual events. They evoke profound emotions in those who observe them." There is some truth in that. Astronomer Serge Brunier describes his experience with a perfect Solar eclipse in his book, *Glorious Eclipses: Their Past, Present, and Future*: "The sight is so staggering, so ethereal, and so enchanting that tears come to everyone's eyes. It is perfect a moment."[4]

Astronomers Guillermo Gonzalez and Jay W. Richards consider perfect Solar eclipses an example of fine tuning. "Besides their intrinsic beauty, perfect Solar eclipses have played an important role in scientific discovery. In particular, they have helped reveal the nature of stars, provided a natural experiment for testing Einstein's General Theory of Relativity, and allowed us to measure the slowdown of Earth's rotation."[5] Gonzales and Richards continued, "....it is hard to exaggerate the significance of the insights afforded by the 1868 and 1879 eclipses for developing stellar astrophysics late in the nineteenth and twentieth centuries."[6]

Yes, we humans are fortunate to have perfect Solar eclipses occurring in our time. Indeed, they have helped to drive the progress of science, especially astrophysics and cosmology, by several decades. However, in my opinion, there is an even fundamentally greater significance of perfect Solar eclipses than just the impact on scientific progress. A phenomenon of perfect Solar eclipses is not "a strange coincidence." It is a perfect example of local fine tuning. *Perfect Solar eclipses carry a profound message to us: "You, humans, have arrived reasonably on time!"*

Appendix 4 presents details of a simplified analysis of perfect Solar eclipses. Based on the data of Appendix 4, I derived Figure 6 which shows a relative frequency of occurrence of perfect Solar eclipses as a function of time (in millions of years).

As Figure 6 shows, perfect Solar eclipses began approximately 610 million years ago, reached a peak in frequency of appearances 220 million years ago, and will end 330 million years in the future. In our epoch, the frequency of perfect Solar eclipses is approximately seven hundred for every thousand years. At their peak 220 million years ago, perfect Solar eclipses occurred 5 to 10 times more often than in our time.

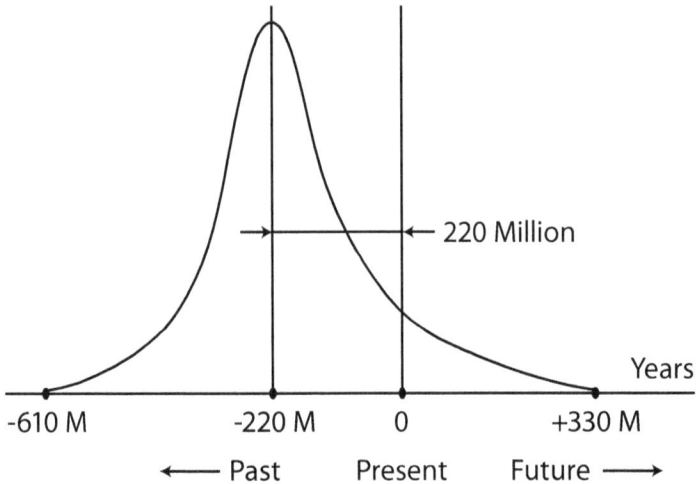

Figure 6. Relative frequency of the occurrence of
perfect Solar eclipses

The data shows that our arrival on the cosmic stage was within the era of perfect Solar eclipses, but we were late by 220 million years (damn those dinosaurs!). It is difficult to escape the impression that the timing of human arrival was pre-set during the formation of the Earth-Moon system 4.5 billion years ago. We humans "achieved" our target within 4.9%. This is outstanding performance taking into account that our planet went through many catastrophes and calamities such as global ice ages, mass extinctions, and the occasional devastating impacts of large asteroids. Still, our score is not perfect. Our rating is B+ at best.

A reader might ask, "Who pre-set the timing?" My answer is that we humans do not know yet. At the present, science has no explanation.

Case Two – Space Energy Density vs. Matter Energy Density

On the largest scale, our Universe is uniform. Matter and radiation are distributed uniformly. The laws of nature are the same everywhere. Our Universe keeps expanding. The total average energy density d_T is a sum of two—and only two—components, d_M and d_S:

$$d_T = d_M + d_S$$

where d_M is matter energy density consisting of ordinary matter, radiation and dark matter (all posses an attractive gravitational property), and d_S is an intrinsic space (vacuum) energy density which is uniform throughout the space and the source of repulsive force causing the Universe to expand.

Space energy density was originally introduced by Einstein in 1917. Space energy density does not change its value with time. It is called a cosmological constant. Einstein's discovery of space energy density was a profoundly intuitive scientific achievement for the wrong reason. Einstein considered it his "biggest blunder." Many scientists would be happy to have such blunders. The cosmological constant should be called the Einstein constant.

According to cosmologist Alex Vilenkin,[7] immediately after the Big Bang, matter energy density d_M exceeded space energy density by many orders of magnitude. For example, at one second after the Big Bang, matter energy density was 10^{45} times greater than space energy density. With the continued expansion of our Universe, matter energy density was diluted and at the present time is approximately equal to space energy density. In the distant future, with the continued expansion of the Universe, the pattern will be reversed. For example, in 100 billion years, matter energy density will be extremely small. It will be 10^{45} times smaller than space energy density.

Vilenkin made the following observation in his book *Many Worlds in One*, "Thus, throughout most of the history of the universe the density of matter is strikingly different from that of the vacuum. Why, then do we happen to live at the very special epoch when the two densities are close to each other? Considering the huge range of variation of the matter density, the coincidence is so extraordinary that it's very hard to dismiss it as 'only a coincidence.'"[8]

Figure 7 shows an example of remarkable fine tuning at the point when d_M reaches equality with d_S. The equality $d_M = d_S$ occurred about 5 billion years ago. It was the crossing point when our Universe passed from matter energy domination to space energy domination. We cannot escape noticing a parallel with perfect Solar eclipses by recalling the equality of two projected images, the Moon and the Sun. Except, in the case of equality of energy densities we are dealing with the whole Universe. The crossing point where two energy densities are equal is thus a universal event. Is there any significance to the crossing point? It must have significance but only for an embryonic-stage intelligence. For an advanced intelligence it would be a trivial issue.

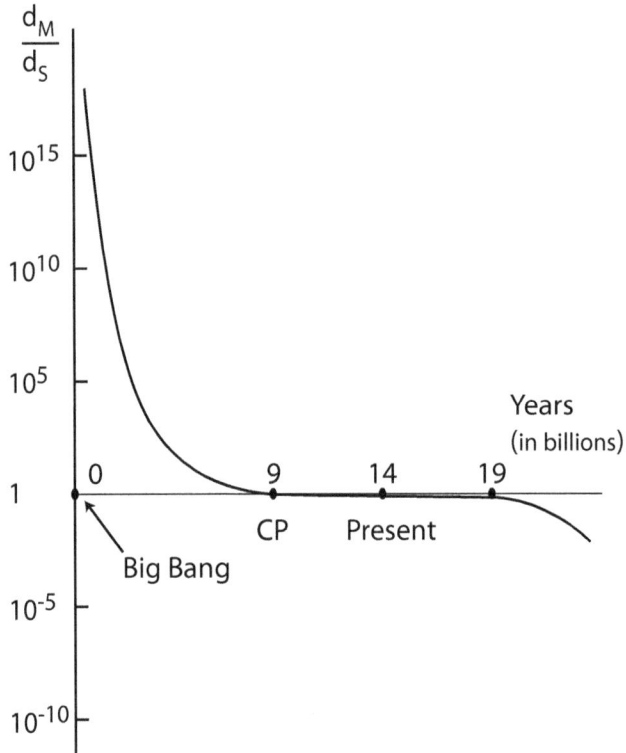

Figure 7. Cosmological evolution of d_M/d_S (CP is the crossing point where $d_M=d_S$)

I dare to offer the following explanation of the significance of the crossing point:

The crossing point at which two energy densities are equal is the moment in the history of our Universe when, for the first time, approximately 5 billion years ago, the evolution of life reached the level of embryonic-stage intelligence on some habitable planets located randomly throughout our Universe.

How does our planet fit into this picture? Overwhelming odds are that we are among the majority entering the cosmic stage. If we are in the middle of the distribution of embryonic-stage intelligence emergence and, assuming that the distribution is symmetrical, then the last group of embryonic intelligence would be entering the cosmic stage about five billion years from now. A conceptual distribution shown in Figure 8 defines *the time window* of ten billion years where one can find embryonic-stage intelligence within, and none outside.

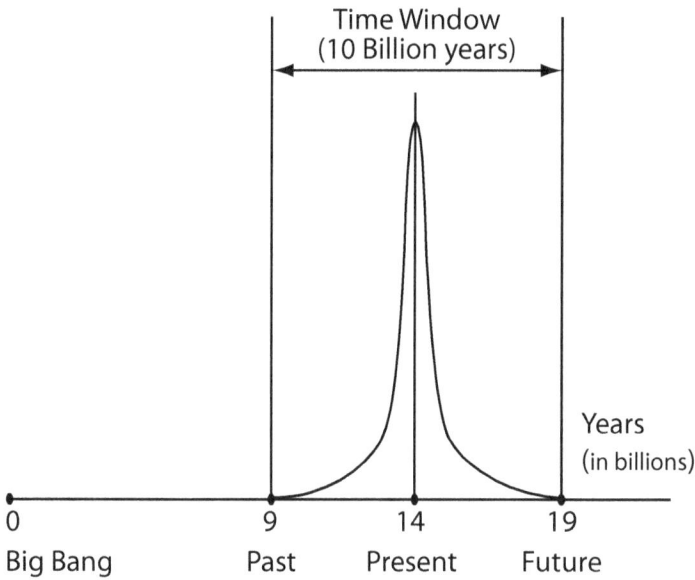

Figure 8. Appearance of embryonic-stage intelligence on the cosmic scene within the time window

We humans are allocated approximately 4 billion years to reach an advanced intelligence level before our situation on Earth will deteriorate due to either (or to both) of two well-known coming events:

- Our Sun will be entering the red giant stage in 4 billion years

- The galactic habitable zone of the Milky Way will experience either some distortion due to a near-miss collision with the Andromeda galaxy, or total destruction in the case of a head-to-head collision, also in 4 billion years

Here is another example of fine tuning as related to our planet Earth—both events have approximately the same time frame. It is a "strange" coincidence.

Using the above analysis, it is possible to estimate the remaining life of our Universe as follows:

- 5 billion years until the emergence of the last group of embryonic-stage intelligence

- 4 billion years for the last group of an embryonic-stage intelligence to reach the advanced intelligence level, and

- An additional 5 billion years as a contingency.

In approximately 14 billion years our Universe would have completed its mission. It would be ready to pass into oblivion. We are at the middle point of the existence of our Universe. That in itself is a "strange" coincidence.

The second law of thermodynamics works relentlessly. The era of the formation of spiral galaxies is over. Spiral galaxies are crumbling. Spiral arm structures of spiral galaxies are getting more and more difficult to maintain, principally for two reasons: either due to influences, collisions or mergers with neighboring cosmic structures (as demonstrated by the Andromeda-Milky Way collision scenario); or due to a shortage of feeding material such as dwarf galaxies, molecular clouds or globular clusters. Without spiral galaxies there can be no galactic habitable zones with habitable planetary systems.

The crossing point $d_M = d_S$ is the remarkable fine tuning factor which allows one to arrive at the following conclusions:

1. Embryonic-stage intelligence appeared on the cosmic scene for the first time approximately 5 billion years ago. Science in the future will determine the timing more accurately.
2. The time window where one can find embryonic-stage intelligences is approximately 10 billion years. We humans appear to be in the middle of the time window.

3. Our Universe will fulfill its purpose and in approximately 14 billion years will pass into oblivion through a natural process unknown to science at the present.

Case Three – The Size of Our Universe

The fact that we still do not know the size of our Universe is an impediment to scientific progress, especially cosmology, physics and philosophy.

Based on the Law of Fine Tuning, I propose a simple estimate which even in its simplicity, is still superior to labeling the size of our Universe as infinite. Even if I am somewhat off in my estimate from the actual size, it still would be "infinitely" more accurate than stating that the size of the Universe is infinite. *It is important to keep in mind that the issue of the size of our Universe is important only to embryonic-stage intelligence and that embryonic-stage intelligence can be found only within the time window.*

We know that an average galaxy has approximately 200 billion stars. We also know that in our observable Universe, there are 200 hundred billion galaxies. In our estimates we can neglect the contribution of some other cosmic entities such as dwarf galaxies, molecular clouds, and globular clusters.

Let's make a crucial observation that the matter contents in our observable Universe change very slowly within the time window of 10 billion years. I then call the matter contents of our observable Universe *the element* of our Universe.

We are now ready to proceed with the final extrapolation, going from 200 billion stars in an average galaxy to 200 billion galaxies in a single element to 200 billion elements in our Universe. That is how I am applying the Law of Fine Tuning.

Then, according to this proposed approach, our Universe has
$$4 \times 10^{22} \text{ galaxies.}$$

So far, all my discussions, assumptions, and estimates have been based on Earth time reference. One has to bring into consideration a universal time reference and raise the issue of synchronization of clocks among habitable planets.

However, the actual synchronization of clocks among habitable planets requires an exchange of light signals. There is no way one can synchronize clocks throughout the Universe because of the huge space separation among habitable systems, habitable galaxies and the elements, and speed of expansion which, depending on separation, can exceed the velocity of light.

There is a way out however. The most important fact is that in the low gravitational environment where all habitable planets reside, the time reference and time speed are the same for each habitable planet. Each individual intelligence counts its time from the moment of our Universe's origin - the Big Bang. In the low gravitational environment, the time is universal throughout our Universe because space expansion is uniform throughout our Universe and no one location is exclusively unique.

23

Our Place in the Universe

Are we alone in the Universe? The question has been asked many times by many scientists and philosophers over the past several centuries. How can we answer it, or even tentatively approach in answering it? Rather than discussing this topic in the form of science fiction, raising all kinds of fantastic scenarios, such as looking for different forms of life based not on carbon but on silicon, or other exotic forms of life in super-hot or super-cold environments, or exoteric life in the great depths of the ocean, let us take exactly the opposite approach by assuming that life as we know it on our planet Earth is the role model and reference for life everywhere in our Universe. In my view, it is a reasonable assumption, for the laws of nature are universal and the same in all parts of our Universe.

Some say that our Universe is bio-friendly. I say, on the contrary, our Universe is extremely hostile to life, but is not in conflict with it. Massive fine tuning was required *locally* to make our planetary system habitable and our planet Earth an exceptional place for the origin and evolution of life.

According to Hugh Ross,[9] this is a partial list of fine-tuned features observed to make our planetary system habitable:

Planetary system *[Solar system]*	137
Star *[Sun]*	140
Moon *[our Moon]*	27
Planet *[Earth]*	268
Planet's surface *[Earth's surface]*	137

The list is not final yet. New fine-tuned features are expected to be added in the future. The numbers are staggering and cannot be explained by coincidence.

Is there anything unique in our cosmic situation? Yes, indeed, but not *exclusively unique.*

The starting point of my discussion is our Milky Way galaxy and its cosmic environment, the Local Group. As pointed out correctly by Hugh Ross,[10] the Milky Way is an exceptional spiral galaxy. It resembles a perfectly-tuned musical instrument. Moreover, it is located in a favorable cosmic location, a quiet corner known as the Local Group. The Local Group has a sufficient number of dwarf galaxies, some of which are being gradually consumed by the Milky Way to sustain its spiral arm structure.

The next big perturbation to the Milky Way galaxy is going to happen in approximately 4 billion years when the Andromeda galaxy will begin merging with the Milky Way, eventually forming one giant elliptic galaxy.[10] By that time, the evolution of life on Earth will be complete, the Sun will be expanding into a red giant, and the Earth will melt away, becoming an ionized cloud and joining the upper layers of the red giant, our former Sun.

Our planetary system is situated in a galactic habitable zone of the Milky Way, located at a radius of 27,000 light years from the galactic center. It is a co-rotation distance, which means that our planetary system revolves around the galactic center at approximately the same rate as the spiral arm structure does, thus avoiding frequent crossings of the spiral arms. Such crossings result in the potential risk that the Solar system might run into a dense dust cloud, remnants of a supernova. This has happened in the past. The next spiral arm crossing by our Solar system is expected in a few hundred million years. As you can see, our habitable Solar system is not guaranteed a risk-free existence. (There is much more to the story but it is not the subject of this book.)

The question can be asked, if the Milky Way's galactic habitable zone is so favorable for life, how many more habitable planetary systems could it accommodate? Progressive creationists say just one. But why waste such exceptional space in the galactic habitable nursery? The circumference of the habitable zone is approximately 170,000 light years. If we allocate conservatively an average separation of 3,500 light years between two neighboring habitable planetary systems, then 50 such systems can be accommodated. Why do we need to have such separation between habitable planetary systems at all? The Law of Fine Tuning requires that close contact between habitable planetary systems, where the evolution of life proceeds independently, must be prevented. How many spiral galaxies with characteristics such as the Milky Way galaxy, with its cosmic environment such as the Local Group, are in existence in the observable Universe? Creationists answer, "None, for God created the Universe just for the benefit of humans and nobody else."

At this stage, astronomers do not have the technological means to answer how many qualified spiral galaxies exist in the observable Universe. As a first step, let us examine the Andromeda, our neighbor galaxy located at a distance of 2.5 million light years from the Milky Way. In the past, Andromeda collided with a medium-sized galaxy. A merger of the two occurred, resulting in the distortion of their galactic habitable zones and the appearance of two black holes in the center. Therefore, the Andromeda galaxy would not qualify to be a nursery of habitable planetary systems.

As mentioned already, Andromeda and Milky Way are on a collision course with an estimated speed of 120 km/second. One could wonder why we need to be concerned with the collision of two galaxies. It is fairly obvious that during such collision, head-to-head collisions of stars are unlikely. However, if a merger or collision of two galaxies were to occur during the cycle of evolution of life, the continuation of the evolution of life would be at risk due to a resulting distortion of a galactic habitable zone. Habitable planetary systems might then find themselves in not-so-friendly locations, such as within a dense galactic center, with all the risks involved.

Returning to the question of how many habitable spiral galaxies exist in the observable Universe, I can put forward a not unreasonable assumption that one out of one million galaxies is a qualified candidate for local fine tuning and, as a result, qualified to be habitable. If such is the case, then out of 200 billion galaxies in the observable Universe, there would be 200,000 habitable galaxies. If we further assume that each habitable spiral galaxy contains fifty habitable planetary systems in its galactic habitable zone, then in our observable Universe the total number of habitable planets would be *ten million*.

As estimated previously (see Chapter "Application of the Law of Fine Tuning in Cosmology: Study of Three Cases), our Universe contains 4×10^{22} galaxies. Then we arrive at the staggering number of 2×10^{18} habitable planetary systems in our Universe. Not all of the habitable planetary systems would be expected to survive though, due to such occurrences as cosmic catastrophes or self-destruction.

I am confident that you, the reader, understand that the estimate of ten million habitable planetary systems in the observable Universe is a speculation on my part. It is based on an arbitrary assumption that one galaxy out of a million is qualified to be habitable, which translates to one star out of 4×10^{16} being qualified to be the center of a habitable planetary system. I consider this assumption reasonable and not in conflict with the Principle of Proportionality.

24

Entanglement and the Origin of Our Universe

According to materialist cosmology, our Universe originated from primordial chaos. The authors Paul J. Steinhardt and Neil Turok said in their book, *Endless Universe*, "Cosmologists differ on the precise properties of this starting state [of our Universe], but many believe it would have been wildly turbulent and non-uniform, with huge variations in density and temperature from place to place, and with space curved and warped in unpredictable ways."[11] The statement about primordial chaos is a surprise because such thinking is reminiscent of philosophers of ancient times, and shows very little progress.

But in our time, such a worldview creates a contradiction: on the one hand, our Universe evolves toward a high value of entropy in accordance with the second law of thermodynamics; on the other hand, our Universe originated from primordial chaos at the highest level of entropy and is evolving toward a high level of entropy, which is nonsense.

Before we continue on this subject, let's streamline the concept of entropy. The second law of thermodynamics and the concept of entropy are outstanding achievements of 19th Century science. In simple terms, entropy reflects a degree of disorder in thermodynamically-isolated systems:

- High organization (high order) means low entropy

- Low organization (low order) means high entropy

It is an inverted way of thinking. To streamline the concept of entropy, I dare to introduce the concept of intropy as *inverted entropy*:

$$intropy = 1/entropy,$$

which means that

- High organization (high order) means high intropy

- Low organization (low order) means low intropy

In a nutshell, *entropy* is a measure of disorder and *intropy* is a measure of order.

In contrast to the materialist concept that our Universe originated from primordial chaos, one should like to think that our Universe originated at the highest level of intropy, having acquired "a huge supply for a long journey." Since its origin, our Universe has been sliding downhill on the intropy slope, gradually spending its supply. Let's assume that our Universe has a negligible energy interaction with the outside world and is thermodynamically isolated in Uni-Universe. In addition, let's also assume that our Universe was in the "absolute" organization at the moment of its origin. Such a thought makes some scientists uncomfortable—it smacks of creationism. But as scientists, we have no choice but to face objective reality as it is.

Creationists have enough trouble with one universe. The thought that in addition to our Universe, there are trillions of other habitable universes puts creationists off-balance. They would wonder: if God created the world for the benefit of humans and only humans, then what was he thinking by creating trillions of other habitable universes?

When I say that Universe was at an "absolute" level of organization, I mean that the organization was beyond human comprehension with deviation from absolute as follows:

$$|absolute - \text{``}absolute\text{''}| \leq 10^{-50},$$

where the assumed factor of 10^{-50} looks "reasonable," meaning "beyond comprehension."

Energy density was "absolutely" uniform; temperature was at the "absolute" zero level; and space geometry was "absolutely" flat. If that is the case, we do not need inflation to justify the current state of our Universe. We do not know how long a pre-Big Bang Universe stayed in such an "absolute" state. Then, some unknown triggering mechanism brought our Universe to the moment of the Big Bang explosion.

However, there is one property which seemingly requires inflation. It is *entanglement*. Here is an example of the intrinsic fundamental interrelationship between quantum mechanics and cosmology. Our Universe, prior to the Big Bang, was in an "absolute" state of entanglement, meaning that each part was entangled with any other part. It would require a giant burst of space expansion, with speed exceeding the velocity of light to break the entanglement and free all parts from each other.

Inflation is not required to explain flatness, uniformity, homogeneity or isotropy of our Universe. These are trivial details. The role of inflation is to break the primordial entanglement and assure long-range separability of individual parts of our Universe. That would seem to be in conflict with quantum mechanics which states that once two systems S1 and S2 interacted in the past and then are separated by any distance, they are entangled. On this issue quantum mechanics is wrong - it does not take into account space expansion, especially inflationary space expansion.

Einstein insisted on separability. He said, "But on one supposition we should, in my opinion, absolutely hold fast: the real factual situation of the system S2, is independent of what is done with system S1 which is spatially separated from the former."[12] He was right only with respect to the long-range separability, which was achieved through inflationary space expansion. Our Universe does not have *universal connections*.

25

Cosmic Seed and Our Universe

The Pre-Big Bang state (the PBB state) of our Universe is not accessible to present-day science. The most important aspect of the PBB state is the packaging of laws of nature, unique to our Universe with all its physical constants already discovered and expected to be discovered in the future, and with all universal fine tuning factors. I call the PBB state the Cosmic Seed. It is there, in the Cosmic Seed, that the deepest secrets of nature are hidden. Again, some scientists would say that it smacks of creationism. Following such logic, one could claim that placing a seed of a plant into moist soil is also an act of creation. A seed is packed with highly organized genetic information in a high *intropy* state. When the seed is placed into moist soil it begins to grow, producing a plant. In a certain sense, it is an act of creation. It is also a miracle.

Similarly, in Uni-Universe, cosmic seeds are packed with superbly organized information, which defines the individual profile of future universes, their origins, evolutions and their eventual disappearance out of existence. All cosmic seeds are organized at the highest level of intropy. Cosmic seeds are "absolutely" fine-tuned to perfection beyond human comprehension.

Is there a purpose to our Universe? Some scientists with a materialistic mentality say that there is no such thing as a purpose to our Universe—the existence of our Universe and our own existence are pointless. I state that there is a purpose to our Universe. The purpose is to originate life in some selected, locally fine-tuned cosmic spots within a specific time frame, and then to evolve life to the advanced level of intelligence.

So far, cosmologists are focused on the simple aspects of the origins of our Universe: space geometry, energy density, and temperature distribution. These are relatively trivial details. Alan Guth thinks that it is possible to produce a universe from a small chunk of matter by bringing it to a super-high density (he called it the "ultimate free lunch"). In my opinion, it would not work without including the laws of nature in the package, which is an impossibility. In addition, one has to place the chunk into the "absolute" state, which is also an impossibility.

In Uni-Universe each individual universe is originated from its individual cosmic seed, thus acquiring its unique "personality." Our Universe is probably a relatively new "design." It is not very efficient. The design itself is seemingly undergoing a process of improvements and refinements via many, many cycles.

The concept of the Cosmic Seed has to be understood with the recognition of the fact that consciousness is a property of objective reality.

In the future, consciousness will be an important part of scientific studies by physicists and cosmologists. Future science has no choice but to cross the line it has been reluctant to cross so far.

Conclusion

When a theoretical physicist talks about the laws of nature, he or she, in fact, means the laws of physics. A theoretical physicist, like the rest of us, knows very little about the laws of nature. The predominant opinion among theoretical physicists is that all processes in nature can eventually be reduced to physics. This is a materialistic point of view, it is wrong. Materialism has different names, such as dialectical materialism, scientific materialism, naturalistic materialism, or materialistic naturalism. It is the most entrenched philosophy in science and is defended with religious intensity.

In my view, it is most important for a scientist/philosopher to follow Einstein and stay on the position of objective reality. Objective reality might have some features or properties not to our liking or expectations but we have no choice but to recognize and accept these features as something beyond our control or even comprehension.

There is a long list of philosophical issues where materialism is wrong, including the following most obvious ones:

- According to materialism, the Universe originated from primordial chaos.

- Materialism is unable to explain why the Universe is habitable.

- Multiverse is an extreme materialist theory which claims the existence of an infinite number of universes each with random laws of nature.

- According to materialism, the Universe has no purpose; the Universe is pointless and our human existence is pointless as well.

- Materialism rejects the Law of Fine Tuning, both universal and local. According to materialism, all can be explained by coincidences regardless of extremely small odds.

The biggest misconception of materialism is its concept of consciousness. According to materialism, consciousness is a product of matter, which I would translate as "consciousness is a product of mindless physical forces". Here one can see a fundamental contradiction and absurdity.

One of the most serious problems for materialism is its denial of local fine tuning (see the Chapter "The Law of Fine Tuning"). After the origin of the Universe, a *local* fine tuning force gets into the action, which cannot be explained by materialism. As I have discussed before, local fine tuning creates a habitable cosmic environment in *qualified locations* throughout our Universe for the origin of life and its evolution. The overwhelming majority of physical processes in our Universe are not affected by local fine tuning. They are *on autopilot,* "confirming" the materialist point of view. In this book I intentionally stay away from life science issues which cannot be explained by materialism such as origin of life, actual mechanism of life evolution, directivity of life evolution, metabolism of a living cell, and so on.

What is an opponent to materialism? The obvious answer is creationism.

What is an opponent to creationism? Another obvious answer is materialism.

Both obvious answers are obviously wrong. Both materialism and creationism have the same opponent. Both are in conflict with the *progress of science.*

I want to emphasize again that materialism cannot explain the origin of the Universe, the origin of the laws of nature and the universal fine tuning. Where did the laws of physics as a subset of the laws of nature come from? The laws did not just drop out of nowhere. In my view, the laws of physics, with their set of physical constants, were preset and fine-tuned to the "absolute" degree prior to the origin of our Universe. After the Big Bang, the physical processes were set *on autopilot.* Here is the justification for the materialist worldview: physical processes are governed by the same laws of physics regardless of where they take place, either in a research laboratory, or in a nearby cosmos, or in any other place of our Universe. This is the only area where materialism can still survive and flourish for the time being.

In the case of creationism, the gap between a body of scientific knowledge and creationism's frozen dogmas is ever-increasing. The future of creationism does not look assured. It is unlikely that creationists would be willing to make their dogmas adjustable with the progress of fundamental science.

Let us recall again the historic episode of Ptolemy/Aristotle model of the world. According to the model, the Earth is the stationary center of the world. This was a very comfortable model for the Catholic Church. "The loving and caring Creator" created the world just for the benefit of humans and placed humans in the center. What had demolished such a beautiful model? It was science, especially astronomy. And it will happen again with the frozen dogmas of creationism. No hand of materialism is needed.

We humans know very little about the laws of nature and objective reality. Most scientific discoveries are still ahead of us.

Both materialism and creationism, with their fixed and inflexible dogmas, have established relatively minor positions in a huge space of possible philosophical systems. The space is wide open and awaiting an advanced philosophy such as ConsReality (see Reference 2 in the Introduction).

The question could be asked why Super Quantum Mechanics is called a super theory. There are at least two explanations:

1. Super Quantum Mechanics is the next step in the progression of quantum mechanics toward a deeper physics theory.

2. Super human effort is required on the part of some scientists/philosophers to abandon and reject their mindset, which is based on a lethal combination of materialism and positivism.

Super Quantum Mechanics is based on a new paradigm, described and explained in Reference 1 of the Introduction. All enigmas, mysteries and paradoxes are melted away. The new paradigm brings ontology into quantum mechanics. The centerpiece of Super Quantum Mechanics is an elementary quantum entity/event.

Super Quantum Mechanics brings consciousness into focus and into a system. Mathematics is a universal language and, surprisingly, properties of consciousness can be described and explained in terms of mathematical formalism.

The philosophy of Super Quantum Mechanics is non-local realism. Super Quantum Mechanics is a searchlight into the deeper level of the quantum world.

Appendices

Appendix 1. Science á la Cro-Magnon

Published literature on the interpretations of quantum mechanics is full of misconceptions and absurdities. Scientists are reluctant to criticize their peers; doing so just does not pay off.

Misconceptions, absurdities and superstitions in science are a well-known phenomenon. Thirty-five thousand years ago, during the Cro-Magnon period, misconceptions, absurdities, and superstitions were already in existence.

The following is a story about one Cro-Magnon tribe of 51 people. The Cro-Magnons lived in a large and comfortable cave. In the late evening, after a tiring and dangerous day of hunting, the Cro-Magnons sat around the bonfire with plenty of time to reflect upon and discuss some "scientific" topics. One phenomenon especially intrigued them. They knew everything about rocks and stones but they could not understand why, when someone throws a stone straight up, the stone somehow returns straight down. The Cro-Magnons had various hypotheses and theories. Gradually, the consensus revolved around one theory: stones enjoy flying up and want to promptly return back into the hands of the thrower for an encore.

One prominent Cro-Magnon did not share such a theory. His name was Cro-Einstein. He enjoyed high prestige and fame among the enlightened members of the Cro-Magnon community. He was famous for his theory where he proved that two small stones are equal in size to a medium stone, and two medium stones are equal in size to a large one. He was skeptical about the evolving consensus view but the debating group did not have enough patience to listen to him. The group decided to proceed with an experiment. What would happen if the stone thrower hid while the stone was still going up? After all, if the stone does not see the thrower while flying up, then perhaps the stone would have no incentive to return. If the experiment proved true, it could be a huge discovery with many promising implications.

So, the Cro-Magnons performed the experiment, many, many times, under strict experimental conditions: the thrower ran fast to hide from view in a cave, or behind a tree, or even behind a giant rock. But to no avail; the stone persistently returned each time to the same spot.

Debates and analyses followed; no satisfactory explanation was found. Along the way, someone produced an original and insightful observation: the thrower was not hiding quickly enough, and when the stone was returning, it could see that the thrower was still hiding. Thus, a new proposal gradually emerged: one should place a large animal skin under the feet of the thrower. After throwing the stone up and while it is still flying up, the thrower would need only an instant to hide under the skin. There was no way the stone would notice his disappearance.

On a side note: doesn't this remind you, the reader, of "delayed choice" to outsmart photons? Or remind you of all those theories on contrived "conspiracies" among the various parts of experimental apparatus for measurement of non-local influences?

When the debating group asked Cro-Einstein his opinion about their new proposal, he answered that although the theory under discussion was quite compelling, his inner voice told him that the theory was incomplete. He felt that a stone is utterly indifferent to the thrower and in fact, is totally oblivious to the thrower's existence. Cro-Einstein believed that there must exist some kind of affinity between the stone and the ground. It appears that stones prefer to stay on the ground rather than to fly straight up forever. Again, nobody had the patience to listen to Cro-Einstein. All were excited: they had in their hands a new hot idea and intended to proceed with the experiment as soon as the hunting season was over.

Cro-Einstein returned to his corner of the cave. There on the wall were many strange symbols which he had previously scratched into the rock. No other Cro-Magnons could make much sense of the symbols. But tonight, still thinking about the stones, Cro-Einstein decided to work on something new and exciting—his own theory of conservation of stones. There had been stories going around the Cro-Magnon tribe about some stones mysteriously appearing or disappearing. Even today he heard a story that a large stone close to the cave entrance disappeared during the night and in the morning no one could find it. Several days before, he'd also heard another story that during the night, one stone had even landed inside the cave with a loud thud; surprisingly, Cro-Einstein had not heard the thud.

All those stories and rumors were persistent. Something must be done to resolve the mystery. So Cro-Einstein arrived at a simple and beautiful idea. He would hide some stones in a specially prepared hole in the ground, wrapping all of them thoroughly and separately in a skin. He would assign to each stone a specific finger on his hands. He would check on the stones again before the next hunting season, and if any were missing, it would not be difficult to identify the missing stone and then look for the explanation of its disappearance. Again, his inner voice told him that stones cannot simply appear or disappear suddenly and mysteriously; to Cro-Einstein, such stories seemed to be absurdities.

Appendix 2. Lecture on The Many Worlds Interpretation

My story goes like this:

A prominent scientist, who has spent many years of his professional life studying and working on the Many Worlds Interpretation, arrives at a university to give a two-hour lecture to students and faculty. In the auditorium are undergraduates, graduates, PhDs and faculty members waiting with great anticipation. The Many Worlds Interpretation is a difficult subject. Few of those attending the lecture have had the patience to study it thoroughly. Many Worlds is the kind of subject that the more you study, the less you understand. The audience is excited for the opportunity to listen to someone who has spent many years of his professional life on the subject and published numerous scientific articles and many books. He is considered the obvious authority and, better still, he possesses a forceful and enthusiastic personality and is an exciting speaker.

The speaker begins by mentioning Hugh Everett, III, a former student of John Wheeler, who in 1957 wrote his PhD thesis titled "On the Foundation of Quantum Mechanics." John Wheeler was so impressed with Everett's theory that Wheeler himself even traveled to see the famous Niels Bohr to discuss the theory and obtain Bohr's opinion. The speaker, however, neglects to mention that Niels Bohr was skeptical of Many Worlds. Bohr had heard many bizarre physics theories in his lifetime, but the Many Worlds Interpretation was the most bizarre.

The lecturer continues: *In his thesis, Everett assumes that a global wave function in Hilbert space provides a complete and accurate description of the state of the entire universe and that this global wave function under all circumstances evolves according to deterministic linear dynamics. There is no such thing as wave function collapse. Observers are modeled as physical systems. Each human is presented as a massive set of quantum states.*

If a measurement is made, every possible quantum state is realized through universe branching, resulting in parallel universes. For example, if an electron recombines with a proton, forming a hydrogen atom, and settles even for a short instance on one of an infinite number of possible atomic energy levels, then the universe branches into an infinite number of parallel universes. How many hydrogen atoms are being formed in our Universe? Zillions! A massive recombination of protons and electrons occurred in the past 380,000 years after the Big Bang, when the Universe was transformed from a plasma state to a transparent state, separating light from matter. Even in our time this process of branching continues at a stupendous pace. Zillions of quantum processes go on inside the sun, the Milky Way galaxy, and cluster and super-cluster galaxies in the observable part of our Universe. All these processes are subject to relentless branching.

The audience is totally flabbergasted and sits in disbelief. But the lecturer projects great confidence and enthusiasm. He has a magnetic personality. He continues: *If you are sitting and watching the famous Schrödinger cat, you are also branching—here you see a live cat, but your copy in another universe is sadly watching a cat's dead body.*

If you are one of the listeners, you are dumbfounded and your head is spinning. It appears that there is no limit to the speaker's unrestrained imagination. In fact, you are told that there are copies of you in an infinite number of parallel universes. The lecturer continues:

You and the infinite copies of you lead separate and independent lives in various parallel universes, including our own. In our Universe you drink coffee; in another universe you read a newspaper; in another, you are still in bed enjoying a late Sunday morning; in yet another you have died in a car accident; in another, you are in a prison cell in a state of profound reflection—you are sentenced to a 25-year prison term for stealing a rare piece of art from a private collection and now you are wondering what went wrong with your thoroughly-prepared master plan; and in yet another universe you are receiving the Nobel Prize for your two-volume fundamental contribution to the Many Worlds Interpretation of Quantum Mechanics.

But the lecturer corrects himself: *I stated that there are an infinite number of copies of you, but it is something of an overstatement. The copying process of an individual is gradually degraded with each split. The idea is to have 10^{100} slightly imperfect copies of oneself, all constantly splitting into further copies that ultimately become unrecognizable.*

As I understand it, your university has an excellent quantum mechanics faculty and a superb quantum optic research laboratory. If you are one of the students studying quantum mechanics, please take a few minutes of your lunch break, go to the quantum optic research laboratory and perform a simple quantum optic experiment using a single photon source and a beam splitter. Direct a single photon on the beam splitter and, voila! You just produced a parallel universe with a copy of yourself and copies of your relatives and friends. Isn't this exciting? And after that you might want to call your parents and tell them about your achievement. Do not hesitate to tell them that you have created a parallel universe in your spare time during the lunch break. I am sure they will be most impressed and might even increase your monthly allowance.

The lecturer then tirelessly answers many questions from the audience. He is superb and sharp. Finally, he asks if anyone still does not understand such a magnificent theory. You know very well that you do not understand it, but you are hesitant to stand up and risk declaring that you are the only one who has failed miserably to see the light. At last, the lecturer announces the end of his presentation. The audience rushes to the exit doors, afraid that the lecturer might change his mind and decide to expand on his presentation. Outside the auditorium, everyone suddenly realizes that, in fact, no one has the slightest idea what the lecture was about.

Appendix 3. Chicks Story

Chicks were living in a yard behind a tall fence on a farm. They were never given an opportunity to see behind the fence. After studying their yard thoroughly for two weeks, the chicks declared that they were ready to write the theory of everything. Except one chick, who, through his effort and intellectual curiosity, uncovered two tiny holes in the fence. The chick peered through the holes and saw something. He hurried to tell his friends that he saw something moving when he looked through the holes. He wondered that perhaps there might be something else that the chicks did not understand or could not even imagine. Maybe, suggested the chick, we should put the theory of everything on hold for a while.

Appendix 4. Details of Perfect Solar Eclipses

Sun (at the present time)

Diameter, DIA	1392000 km
Maximum distance from Earth, APHELON	152 x 10^6 km
Average distance from Earth, AVERAGE	149 x 10^6 km
Minimum distance from Earth, PERIPHELON	147 x 10^6 km

Moon (at the present time)

Diameter, dia	3476 km
Maximum distance from Earth, apogee$_1$	405400 km
Average distance from Earth, average$_1$	385000 km
Minimum distance from Earth, perigee$_1$	362600 km

where apogee$_1$, average$_1$ and perigee$_1$ are values at the present time.

Following are estimates for the era ("the Era") of perfect solar eclipses:

The Era of perfect solar eclipses began 610 million years ago when
$$(DIA/PERIPHELON) = (dia/apogee_2),$$
where apogee$_2$ is the value of apogee 610 million years ago.

Maximum frequency of occurrence of perfect solar eclipses happened 220 million years ago when
$$(DIA/AVERAGE) = (dia/average_3),$$
where average$_3$ is the average distance of Moon from Earth 220 million years ago.

The Era of perfect solar eclipses will end 330 million years in the future when

$$(DIA/APHELON) = (dia/perigee_4),$$

where $perigee_4$ is the value of perigee 330 million years in the future.

The preceding estimates are based on the following assumptions:

- Moon and Sun orbital eccentricities are unchanged during the Era.

- The diameter of the Sun grows 6 cm per year[1] during the Era.

- Annular receding distance "y" of the Moon from Earth is linearly extrapolated to any point in time during the Era using the following assumptions:

  y = 4.8 cm/year 500 million years ago
  y = 3.8 cm/year at the present time[1]
  y = 2.8 cm/year 500 million years in the future.

Appendix 5. My Short Story

After studying all those quantum mechanics interpretations, a typical young scientist, becoming thoroughly confused, gives up, saying "philosophy is useless." The young scientist's conclusion is wrong. Einstein was a great philosopher. His philosophy remains a guiding star for scientists. But philosophy is not a part-time hobby. Philosophy, just like science, music, and art requires passion, energy, talent, and destiny. It also requires intensive professional training. Here I will share with you why I am not a creationist and no longer a materialist. Like the binary code, most people are divided into two groups: creationists or materialists. Of course, each group has many subgroups. I belong to neither of those groups.

As a physicist, I studied dialectical materialism for five years at the Gorki State University in the Soviet Union. I continued studies of dialectical materialism for another three years during my pre-doctoral program at the Joint Institute for Nuclear Research, Dubna, in the Soviet Union.

Dialectical materialism is a well-organized and sophisticated philosophical doctrine. It is convincing. It was the official philosophical doctrine in the Soviet Union and a mandatory program in all universities and colleges. Perhaps it might surprise some people that Soviet students were allowed—and in fact encouraged—to study other philosophical doctrines in their spare time, including the great philosophers of classic Greece and Europe such as Kant and Hegel, which I did with great enthusiasm. Philosophy, the theory of special relativity, and quantum mechanics were my favorite subjects.

Did I believe one hundred percent in dialectical materialism? I should like to think so—very close to one hundred percent. But sometimes I had fleeting anxieties when I looked at the blue sky with small white clouds, looked at wild flowers, or listened to classical music.

The most important postulate in dialectical materialism states that *consciousness is a product of matter.* This postulate is erroneous and harmful to humanity and inhibits scientific progress.

Dialectical materialism, or other branches such as materialist naturalism or scientific materialism, cannot explain the purpose and origin of our Universe, its immense fine tuning, the origin of life, directivity of life evolution, and the evolution of intelligence. In simple terms, materialism cannot explain why we, humans, perceive the sky as blue.

In 1966, as a Soviet scientist, I arrived at the European Center for Nuclear Research (CERN) in Geneva, Switzerland to participate in a joint program for the design and construction of high-energy physics equipment. It was a great opportunity for me. CERN represents a remarkable collection of people of various backgrounds and nationalities. In our research and engineering group we had people from France, Germany, Holland, Switzerland and Italy. Initially, my English was rudimentary and an impediment for communication. Standing before a blackboard with chalk in my hand, I remember trying to explain to my foreign colleagues something about dialectical materialism. There was friendly laughter. It was obvious to them that this Russian guy, who was brainwashed in a communist doctrine, now was spilling his stuff on them. Likewise, I laughed at them, realizing that my new friends were not at all equipped to deal with philosophical issues.

Our group was in charge of a research and engineering project for the design and construction of two large and sophisticated systems to be used in high-energy physics experiments. One system would remain at CERN and the second system would be delivered to the Soviet Union. The head of our group, Dr. Herbert Lengeler, was an individual of great organizational skills and superb professional knowledge. He agreed that in parallel to my duties in his project, I could continue with my own research that I had begun in the Soviet Union. My research included developing a theory of electromagnetic hybrid waves for a special class of cylindrical structures, designing an experimental setup, and performing experiments with the purpose being to compare the theory with the experimental data results. My research was successful and culminated in a doctoral dissertation (thèse d'État), which I wrote in French and submitted to the University of Paris, Orsay, France. In 1970, I presented my thesis before a panel of scientists and students from the University of Paris and subsequently was awarded the Ph.D. It was my personal triumph. And at that time, receiving a Ph.D from a foreign university was an unheard-of event for a Soviet scientist.

References

Introduction

1. Victor Vaguine, "Conceptual and Philosophical Foundations of Super Quantum Mechanics," ConsReality Press, Dallas, Texas (to be published in 2012).
2. Victor Vaguine, "Ontology of ConsReality" (working title), ConsReality Press, Dallas, Texas (scheduled for publication in 2014).
3. Murray Gell-Mann, "What are the Building Blocks of Matter?" in Huff and Prewett, 1979, p. 29.
4. Stephen Hawking and Leonard Mlodinow. "The Grand Design," Bantam Books, New York, 2010, p. 73.

Part I. Quantum Mechanics Issues

1. Abraham Pais, "Subtle is the Lord... The Science and the Life of Albert Einstein." Oxford: Oxford University Press, 1982, p. 9.
2. Murray Gell-Mann, "Questions for the Future" in *The Nature of Matter*, Wolfson College Lectures 1980, ed. J. H. Mulvey, (Clarendon Press, Oxford 1981).
3. Richard Feynman, "The Character of Physical Law," Random House, New York, 1994, p. 123.
4. "The Ghost in the Atom," P.C.W. Davies and J.R. Brown, eds., Cambridge University Press, 1999, pp. 127, 131.
5. John S. Bell, "Speakable and Unspeakable in Quantum Mechanics: Collected Papers on Quantum Philosophy," Cambridge University Press, 1997.

6. Roger Penrose. Foreword to "Quo Vadis Quantum Mechanics," A.C.Elitzur, S.Dolev and N.Kolenda, eds, Springer, 2005.
7. Mark P. Silverman, "Quantum Superposition: Counterintuitive Consequences of Coherence, Entanglement, and Interference," 1st ed., Springer – Verlag Berlin Heidelberg, 2008, p. 26.
8. Thomas S. Kuhn, "Black-Body Theory and the Quantum Discontinuity," Clarendon Press, Oxford, 1978.
9. Brian Greene, "The Fabric of the Cosmos: Space, Time, and the Texture of Reality," Alfred A. Knopf, New York, 2004, p. 112.
10. Werner Heisenberg, "Physics and Philosophy; the Revolution of Modern Science," HarperCollins Publishers, Inc., New York, 1999, p. 160.
11. M. Jammer, "The Philosophy of Quantum Mechanics: The Interpretations of Quantum Mechanics in Historical Perspective," John Wiley & Sons Inc., 1974, p. 161.
12. Ibid., p. 204.
13. John S. Bell: "On the Einstein-Podolsky-Rosen Paradox", Physics **1**, #3 195, (1964).
14. Henry Stapp: "Are Superluminal Connections Necessary?" El Nuovo Cimento, **40B**, 191-205 (1977).
15. S.J. Freedman and J.F. Clauser, "Experimental test of local hidden-variable theories," Phys. Rev. Lett. **28,** 938 (1972).
16. E.S. Fry, and R.C. Thompson, "Experimental Test of Local Hidden-Variable Theories," Phys. Rev. Lett. **37,** 465 (1976).
17. A. Aspect, P. Grangier and G. Roger, "Experimental Realization of Einstein-Podolsky-Rosen-Bohm Gedankenexperiment: A New Violation of Bell's Inequalities," Phys. Rev. Lett. **49,** 91 (1982).
18. Peter Woit, "Not Even Wrong: The Failure of String Theory and the Search for Unity in Physical Law," Basic Books, New York, 2006, p. 203.

19. Brian Greene, "The Fabric of the Cosmos: Space, Time, and the Texture of Reality," Alfred A. Knopf, New York, 2004, p. 115.
20. Daniel F. Styer. "The Strange World of Quantum Mechanics." Cambridge University Press, 2000.
21. Brian Greene, "The Fabric of the Cosmos: Space, Time, and the Texture of Reality," Alfred A. Knopf, New York, 2004, p. 121.
22. Bruce Rosenblum and Fred Kuttner, "Quantum Enigma: Physics Encounters Consciousness," Oxford University Press, New York, 2006.
23. Richard Feynman, "The Character of Physical Law," Random House, New York, 1994 Modern Library Edition, p. 137.
24. Ibid., p. 139.
25. The Feynman Lectures on Physics, vol.3, Addison-Wesley, Reading, 1965, p. 1-1.
26. Stephen Hawking and Leonard Mlodinow. "The Grand Design," Bantam Books, New York, 2010, p. 73.
27. Brian Greene, "The Fabric of the Cosmos: Space, Time, and the Texture of Reality," Alfred A. Knopf, New York, 2004, pp. 119-120.
28. Ibid., p. 121.
29. Stephen Hawking and Leonard Mlodinow. "The Grand Design," Bantam Books, New York, 2010, p. 82.
30. "The Ghost in the Atom," P.C.W. Davies and J.R. Brown, eds., Cambridge University Press, 1999, p. 60.
31. Peter Woit. "Not Even Wrong; The Failure of String Theory and the Search for Unity in Physical Law," Basic Books, New York, 2006.
32. Mark P. Silverman, "Quantum Superposition: Counterintuitive Consequences of Coherence, Entanglement, and Interference," 1st ed., Springer–Verlag Berlin Heidelberg, 2008, p. 25.
33. Murray Gell-Mann. "The Quark and the Jaguar: Adventures in the Simple and the Complex," Freeman and Company, New York, 1994, p. 153.

34. Contemporary Jewish Record 8, an interview with Alfred Stern, 1945, p. 245-9.
35. P.A. Schilpp, ed., "Albert Einstein, Philosopher-Scientist," MJF Books, New York, 1949, p. 87.
36. Ibid., p. 85.

Part II. Philosophical Issues

1. Lee Smolin. "The Trouble with Physics: The Rise of String Theory, the Fall of Science, and What Comes Next," First Mariner Books edition, Houghton Mifflin Company, Boston, 2007.
2. Brian Greene. "The Hidden Reality: Parallel Universes and the Deep Laws of the Cosmos," Alfred Knopf, New York, 2011, p. 309.
3. Ibid., p. 309.
4. Jeremy Bernstein. "Quantum Profiles," Princeton University Press, Princeton: New Jersey, 1991, p. 72.
5. James T. Cushing. "Quantum Mechanics: Historical Contingency and the Copenhagen Hegemony," The University of Chicago Press, Chicago, 1994.
6. Stephen Hawking and Leonard Mlodinow. "The Grand Design," Bantam Books, New York, 2010, p. 118.
7. John Gribbin, "The Origins of the Future; Ten Questions for the Next Ten Years," Yale University Press, New Haven and London, pp. 184-213.
8. Niels Bohr, "Atomic Theory and the Description of Nature," Cambridge University Press, 1934, p. 115.
9. Stephen W. Hawking. "The Theory of Everything: The Origin and Fate of Our Universe," New Millennium Press, 2002, p. 148.

Part III. Current Cosmology and Physics Issues

1. Alan H. Guth. "The Inflationary Universe: The Quest for a New Theory of Cosmic Origin," Perseus Publishing, New York, 1997, p. 186.
2. Paul J. Steinhardt and Neil Turok. "Endless Universe: Beyond the Big Bang," Doubleday, New York, 2007, p. 52.

3. Ibid., p. 91.
4. Andrei Linde, "Inflation, Quantum Cosmology, and the Anthropic Principle," in "Science and Ultimate Reality: Quantum Theory, Cosmology, and Complexity." Eds. John D. Barrow, Paul C. W. Davies and Charles L. Harper, Jr., Cambridge University Press, 2004, pp. 426-458.
5. Lee Smolin. "Trouble with Physics: The Rise of String Theory, the Fall of Science, and What Comes Next," First Mariner Books, Boston, 2007.
6. Peter Woit. "Not Even Wrong; The Failure of String Theory and the Search for Unity in Physical Law," Basic Books, New York, 2006.

Part IV. Advanced Cosmology Ideas

1. Brian Greene. "The Hidden Reality: Parallel Universes and the Deep Laws of the Cosmos," Alfred Knopf, New York, 2011, p. 309.
2. Hugh Ross, "Why the Universe is the Way it is," Bakers Books, Grand Rapids, MI, 2008, p. 50.
3. Ibid., pp. 122-123.
4. Serge Bruinier. "Glorious Eclipses: Their Past, Present, and Future," Cambridge University Press, New York, 2000, p. 17.

5. Guillermo Gonzalez and Jay W. Richards. "The Privileged Planet: How Our Place in the Cosmos is Designed for Discovery," Regnery Publishing, Washington, DC, 2004, p. 10. *Reprinted by special permission of Regnery Publishing Inc., Washington, D.C.*
6. Ibid., p. 15.
7. Alex Vilenkin. "Many Worlds in One: The Search for Other Universes," Hill and Wang, New York, 2006.
8. Ibid., p. 126.
9. Hugh Ross, "Why the Universe is the Way it is," Bakers Books, Grand Rapids, MI, 2008, p. 123.
10. Ibid., pp. 65-69.

11. Paul J. Steinhardt and Neil Turok. "Endless Universe: Beyond the Big Bang," Doubleday, New York, 2007, p. 52.
12. P.A. Schilpp, ed., "Albert Einstein, Philosopher-Scientist," MJF Books, New York, 1949, p. 85.

Appendices

1. Guillermo Gonzalez and Jay W. Richards. "The Privileged Planet: How Our Place in the Cosmos is Designed for Discovery," Regnery Publishing, Washington, DC, 2004, p. 18. *Reprinted by special permission of Regnery Publishing Inc., Washington, D.C.*

Index

About the author

Victor Vaguine is a scientist/philosopher born and educated in Russia (the former USSR). He received his Master's Degree in Radiophysics from the Gorki State University, Russia. From 1966 to 1971, he participated in the joint CERN-USSR program for development of equipment for high energy physics experiments at the European Center for Nuclear Research (CERN), Geneva, Switzerland. He received his Doctorate Degree in Physics (thèse d'État) from the University of Paris, Orsay, France.

After arriving in the United States in 1971, he accepted a position as a scientist at Varian Corporation, Palo Alto, California, an established high technology company specializing in radiation therapy equipment, and in a few years became head of Varian's research & engineering operation. He left his secure position to take an active role in high technology startup companies.

Dr. Vaguine holds 16 United States patents and has published numerous scientific papers. Presently, he is developing his Super Quantum Mechanics theory and his philosophy ConsReality.

This book, *Prologue to Super Quantum Mechanics,* is the first in a series of books. Dr. Vaguine is currently preparing for publication *Conceptual and Philosophical Foundations of Super Quantum Mechanics* as well as *Ontology of ConsReality.*

(For more information about Victor Vaguine, see Appendix 5 "My Short Story").

Dr. Vaguine can be contacted at victorvaguine@gmail.com

www.ingramcontent.com/pod-product-compliance
Lightning Source LLC
Chambersburg PA
CBHW072254210326
41458CB00073B/1711